T0261511

Luminos is the Open Access monograph publishing program
from UC Press. Luminos provides a framework for preserving and
reinvigorating monograph publishing for the future and increases
the reach and visibility of important scholarly work. Titles in
the UC Press Luminos model are published with the same high
standards for selection, peer review, production, and marketing as
those in our traditional program. www.luminosoa.org.

Rivers of the Anthropocene

Rivers of the Anthropocene

Edited by

Jason M. Kelly, Philip V. Scarpino, Helen Berry,
James Syvitski, and Michel Meybeck

UNIVERSITY OF CALIFORNIA PRESS

University of California Press, one of the most distinguished university presses in the United States, enriches lives around the world by advancing scholarship in the humanities, social sciences, and natural sciences. Its activities are supported by the UC Press Foundation and by philanthropic contributions from individuals and institutions. For more information, visit www.ucpress.edu.

University of California Press
Oakland, California

Suggested citation: Kelly, J. M., Scarpino, P. V., Berry, H., Syvitski, J., and Meybeck, M. *Rivers of the Anthropocene*. Oakland: University of California Press, 2018. doi: https://doi.org/10.1525/luminos.43

Library of Congress Cataloging-in-Publication Data

Names: Kelly, Jason M., editor. | Scarpino, Philip V., editor. | Berry, Helen, 1969- editor. | Syvitski, James P. M., editor. | Meybeck, M. (Michel), editor.
Title: Rivers of the Anthropocene / edited by Jason M. Kelly, Philip Scarpino, Helen Berry, James Syvitski, and Michel Meybeck.
Description: Oakland, California : University of California Press, [2018] | Includes bibliographical references and index. |
Identifiers: LCCN 2017026927 (print) | LCCN 2017030862 (ebook) | ISBN 9780520967939 (ebook) | ISBN 9780520295025 (pbk. : alk. paper)
Subjects: LCSH: Rivers—Environmental aspects. | Human ecology.
Classification: LCC GF63 (ebook) | LCC GF63 .R58 2018 (print) | DDC 551.48/3—dc23
LC record available at https://lccn.loc.gov/2017026927

25   24   23   22   21   20   19   18   17
10   9   8   7   6   5   4   3   2   1

CONTENTS

LIST OF FIGURES

FOREWORD

James Syvitski

Rivers respond to precipitation on the landscape after satisfying the needs of the ground, plants, and atmosphere. The resulting surface runoff merges into drainage channels to form continental networks that then carry nutrients and carbon to support our living planet. On any other Earth-like planet, sans humans, we should see similar drainage networks, organized from small streams that connect to larger rivers—perhaps forming a monster river akin to Earth's Amazon. Most rivers carry a strong seasonal signal within their water levels and transport volumes and thus become an important pulse of our planet. Sediment eroded from highlands and mountains both form the channels themselves and supply the important material mass to floodplains, wetlands, deltas, and oceans.

Some mammals, such as the Canadian beaver, have uncovered an evolutionary advantage in modifying the flow of water through the landscape by building houses and other barriers to slow the seasonal pulses of flow. Not until recently, however, has a single species, *Homo sapiens,* taken command of Earth's surface to the point at which the dynamics normally associated with the natural pulses of energy, fluid, and matter have become fundamentally altered.

Human societies have built one large dam (15+ meters in height) every day, on average, for the past 130+ years. We have diverted river water to secure food and power and even to entertain our ever-increasing population. We are presently adding to our population at a rate of one million persons every five to ten days, and this trend will continue for the foreseeable future, at least the next hundred years. Where rivers once supplied nutrients and sediment to nourish our coastal regions, ever increasing numbers of them now run dry for ever longer periods, among them, the Colorado, Yellow, and Indus Rivers. Our waterways once proffered uninterrupted transportation pathways into our continents, not just for us, but for other mammals as well as aquatic life. Today our rivers are often dissected by dams and other barriers, supporting an increasingly engineered landscape. By building flood embankments along fluvial corridors, we have separated the terrestrial landscape from the rivers. We have fixed rivers in place where they once ran wild over vast floodplains. As a result, our rivers (the Yellow River is a good example) have become super-elevated above the historical floodplain. They have become engineered continuations of our city sewage systems. Some say that we have entered a new geological epoch of our own making, called the Anthropocene, in which the human footprint has reached levels akin to the impact of an ice age.

We have changed the species distribution on our planet, with many species on their way to extinction. Invasive species may hitchhike along our transport pathways (e.g., the Japanese knotweed, *Fallopia japonica;* the freshwater zebra mussel, *Dreissena polymorpha*). And humankind's changes to Earth's environments have not finished. Countries around the world are planning thousands of kilometers of new canals to address the twenty-first-century water crisis. Approximately 1.1 billion people today do not have access to safe drinking water, and 2.6 billion are without adequate sanitation; another 1.7 billion people are living in areas where groundwater is being extracted faster than it can be replenished. Significantly, with massive and exponential growth in human populations, world agriculture accounts for 71 percent of global freshwater use. The giant Ogallala aquifer in the United States once had an average water depth of 240 feet; today it is but 80 feet. With humanity's escalating influence on climate, we are changing the global hydrological cycle, altering the extent of snow cover, permafrost, sea ice, glaciers, and ice caps—all leading to changes in ocean volume. A warming atmosphere holds more water and is leading to an intensification of the hydrological cycle. Wet regions are becoming wetter (more flooding); dry regions, drier. Climate change is already bringing about drought and disease, and will do so at a greater rate in the future. Pollution further limits our already stressed resource base and negatively affects the health of aquatic life forms and terrestrial fauna, including human beings. Humans have, so far, achieved water security through short-term and costly engineering solutions. Faced with a choice of water for short-term economic gain or for the general health of aquatic ecosystems, societies through their governments and corporations overwhelmingly choose development, often with deleterious consequences on the very water systems that provide the resource.

I first met Jason Kelly at a 2013 Water Congress sponsored by the Global Water System Project that was held in Bonn, Germany. Jason is a social scientist, and he wanted to give voice to how we got to our present human-impacted river systems. He argued for recognizing the role of rivers in the history of humanity, and what it might mean if rivers were no longer the planetary pulse of our continents. He argued that we must understand how humans think and make decisions, and take nature into account, if we wish our societies to move toward a more sustainable pathway. He proposed that there are two steps: recognize these environmental problems (involving the diagnostic expertise of natural science and engineering), then analyze the conditions that lead to them (social science and humanities). Perhaps at the end of this process, through ongoing transdisciplinary collaboration, it would be possible collectively to turn bad practices and bad decisions around.

Through Jason, I attended the follow-on "Rivers of the Anthropocene" conference in 2014, held in Indianapolis at Indiana University (IUPUI). The conference brought together an even mix of natural and social scientists and scholars from the arts and humanities. Representatives from the Anthropocene Working Group (AWG), a subcommission of the International Commission on Stratigraphy, were also in attendance. The AWG is tasked to determine whether humanity has indeed created conditions on Earth's surface to produce a recognizable global signal in the rock record. I was blown away by the conference. I kept telling people about my experience with the conference participants. In a 2014 interview with the journal *Nature Climate Change,* I noted how exciting it was to see what each academic community could bring to our understanding through the ongoing Rivers of the Anthropocene project and how we had much to learn from one another. I even ventured that perhaps we might look back on this project as laying out a different way to construct higher education, away from the siloing that now defines our academies and universities.

I salute the contributors to this volume for their integrity and scholarship. This is a book for everyone. You can go back to the Great Tyne flood of 1771 and learn of its cause and impact on the community. Perhaps you might discover that river engineering, if done well, can transform a society, such as has been accomplished by the island state of Singapore. Or perhaps you might be intrigued by how artists and scientists can join forces and reveal the water system of the White River in the twenty-first century. These and the many other topics in this volume present a splendid reflection on humans and their interaction with nature. It is a pleasure to write the foreword to this upbeat and insightful book. Thank you, Jason, for pulling me into your approach to our world. Namaste.

James Syvitski
Executive Director, Community Surface Dynamics Modeling System,
University of Colorado at Boulder (USA), and Chair,
International Geosphere-Biosphere Programme (IGBP),
Swedish Academy of Sciences, Stockholm

August 2017

PREFACE

Jason M. Kelly

Humanity is facing a crisis of its own making. The climate is changing. Oceans are warming. Dead zones of hundreds and thousands of square miles hover off our coasts. A mass extinction is in progress—the likes of which have not been seen for 65 million years. Salinization, pollution, and overconsumption threaten our supplies of freshwater. Our environments can no longer absorb human pressures. This is the condition of the Anthropocene—an age in which humans are altering the planet to such an extent that we are leaving a permanent and irreversible mark on its biological, hydrological, atmospheric, and geological systems. Humanity has initiated an environmental "phase shift," and formerly resilient systems have been pushed into altered states. Even if humanity were to significantly modify its behaviors, the result would be a new equilibrium, fundamentally different from that of the preindustrial world.

Identifying and working within environmental boundaries could mitigate the most extreme environmental consequences of human activity, and this is the approach favored by an increasing number of earth systems researchers. However, this will require dramatic shifts in consumption patterns, scientific assumptions, sociopolitical structures, and cultural systems. It will necessitate not only macro-level changes requiring unprecedented transnational cooperation but also micro-level adjustments in the practices of our everyday lives. To state it simply, putting the brakes on runaway environmental devastation will require a whole-sale reworking of our societies, both from a technological-scientific standpoint and from a sociocultural standpoint. Research, planning, and implementation will require close collaboration between experts on the earth's biophysical systems and human sociocultural systems—between scientists, humanists, social scientists, artists, policy experts, and community-based organizations.

Unfortunately, however, in this era in which humans and human systems have become prime agents of changes to the planet, we have yet to create a research and policy culture that bridges the divides between these groups. Because of this, we lose an important tool for tackling some of humanity's biggest issues, detracting from our overall understanding of global ecological change and limiting our ability to respond to escalating crises.

One of the most potentially productive approaches to bridging these divides is transdisciplinarity, an approach that addresses a problem by building research frameworks and methods that transcend disciplinary barriers (Jahn, Bergmann, and Keil 2012; Leavy 2012; Mattor et al. 2013; Palsson et al. 2013; Kelly 2014; Nicolescu 2014). As such, transdisciplinarity is more than simply borrowing methods from other disciplines. As suggested by Jean Piaget (1974, 170), it is a system "without stable boundaries between the disciplines."[1] Building a solid transdisciplinary research structure, however, requires constant and close collaboration between individuals who traditionally work in disciplinary silos. Changing the culture of research—even within a relatively small research cluster—does not happen overnight. It happens only when researchers are willing to question their own epistemological and methodological assumptions while in dialogue with fellow researchers from diverse disciplinary backgrounds. It was in the spirit of transdisciplinary cooperation that the Rivers of the Anthropocene (RoA) project was established in 2013. The mission of RoA is to create an international collaborative network of scientists, social scientists, humanists, artists, policy makers, and community organizers to produce innovative transdisciplinary research on global freshwater systems. These collaborations have resulted in research projects, publicly engaged scholarship, educational outreach, and service work.

The study of global river systems is an ideal arena for developing a transdisciplinary framework for environmental research. Not only is freshwater one of the most pressing concerns of the twenty-first century, but river systems are structures that exemplify the complicated and complex dynamics of human-nature entanglements. RoA starts from the perspective that transdisciplinary approaches are central to understanding the human-environment interface in all its complexities. It is not enough that scientists and engineers measure what humans have done or what they can do to shift environmental processes; it is necessary that they work hand-in-hand with humanists and social scientists to understand the limits and feedback mechanisms that beliefs, practices, ideologies, social structures, and cultural norms impose on human action, which in turn shapes anthropogenic environmental change. Likewise, it is not enough for scholars to analyze the biophysical-sociocultural interface; it is necessary for them to engage in the worlds beyond academia—to work with policy makers, artists, and community organizations to both educate and design better responses to environmental challenges.

The RoA Network is a growing community that currently includes over thirty artists, scientists, humanists, social scientists, policy makers, and community organizers who focus on global river networks during the Age of the Anthropocene. During its first phase, the RoA Network is hosting a series of conferences and workshops focused on developing an integrated, transdisciplinary framework of principles, goals, and methodologies (the RoA Framework) that will offer a working model for interdisciplinary teams of environmental researchers who wish to bridge the divides between the academic disciplines as well as between academic and public policy–oriented work. This book is the product of the RoA Network's first experiments in bridging disciplinary divides. Using freshwater systems as a framing device, the essays in this volume address a series of themes fundamental to examining the intersection of biophysical and human sociocultural systems during the Anthropocene. Consequently, while the authors' primary interests are in water research, the issues with which they engage and the conclusions that they draw echo far beyond the realm of water policy.

. . .

Eighty percent of the world's population is under the imminent threat of water insecurity and biodiversity loss (Rockström et al. 2009).[2] Simply put, water security is one of the most pressing ecological problems of this century. This challenge cannot be solved by creative technological or policy solutions alone. It requires a holistic approach premised on a better understanding of the complex dynamics between human societies and their environments.

Historically, river systems have been central to human societies and their technologies, and these have been of special interest to environmental scholars. Environmental historians, for example, have conclusively shown that rivers are not simply physical landscapes; they are cultural worlds as well—shaped at the interface between humans and nature. These interactions have not always been negative for biological systems. In fact, in some cases, humans have ameliorated some of the more extreme impacts of their activities, allowing their own and other species to flourish. Nevertheless, it is clear that during the Anthropocene humans have had dramatic—and often unintended—negative impacts on river systems. Human-induced salinization, arheism, chemical contamination, and a host of other riverine syndromes can be described and measured through historical data sets (Meybeck 2003). Transformation of river systems through technology such as dams, which regulate two-thirds of the planet's running water, are measurable, contributing to significant transformations of the geomorphology of river deltas and even continental shelves(Syvitski and Kettner 2011). Multiple data sets suggest not only increased anthropogenic changes to the planet during the past 250 years, but dramatic global transformations of earth systems since 1950—a period some term the "Great Acceleration" (Steffen et al. 2004; Steffen, Crutzen, and McNeill 2007; Steffen et al. 2008; Steffen et al. 2011).[3]

Freshwater environments are one of the most vulnerable points in the earth's ecosystem. Currently, farming, mining, industry, and other human processes use half of all the freshwater that exists. Given the fact that humans make up a small fraction of the earth's biomass, we consume far beyond our share. Humans have chemically altered much of the available freshwater—which makes up only 2.8 percent of all the water on the planet—transforming the freshwater cycle and the other biophysical systems that rely on it. In fact, the freshwater cycle is one of nine "planetary life support systems" currently threatened by environmental change, according to Johan Rockström and Will Steffen (Rockström et al. 2009).

Human interactions with their water systems are both amplified and limited by sociocultural motives: culture shapes attitudes, and society determines actions. These attitudes and actions are agents in shaping the planet's organic and non-organic systems. It is clear that the solutions to humanity's water crisis are not simply technological; they are also social, cultural, and political. Therefore, it is essential that specialists from the earth sciences, human sciences, and humanities work together to solve them.

For over a decade, climate scientists, especially those in earth systems science, have been advocating a more interdisciplinary approach to understanding the planet. Earth systems science is predicated on the concept that the earth is a system of complex interactions between the atmosphere, hydrosphere, lithosphere, and biosphere. To understand these interactions, scientists have to pay close attention to biophysical systems—especially anthropogenic biophysical systems, which include human population patterns, sociocultural structures, and political economies. However, most research projects that fall under the rubric of earth systems science either ignore or pay little attention to the complexities of human systems in calibrating their models. Most important, the agency of human groups and individuals gets lost in scientific analysis. This is especially a problem when studying the late Holocene, particularly the period since 1750 C.E. when humans became a dominant force affecting the entire earth system. The inability to integrate human systems into the environmental analysis of the Anthropocene severely hampers technological, educational, and policy responses.

The social sciences and humanities have proceeded along a research track parallel to environmental scientists over the past thirty years. This is especially true in fields such as history, sociology, geography, and anthropology, which have documented the history of humanity's interactions with its environments. More recently, an approach known as environmental humanities has emphasized interdisciplinarity to bridge the divide between the sciences and humanities. In addition to integrating many approaches prominent in the humanities, the environmental humanities have been strongly influenced by science, technology, and society studies, which has not only emphasized that science is embedded in larger sociopolitical structures, but has also articulated the idea that human systems and natural systems are not discrete. The environmental humanities are often characterized by approaches that seek to transcend the descriptive or analytical but wish to have a practical impact, often through the form of criticism. While scholarship in the social sciences and the humanities has come to many similar conclusions as earth systems science, it has tended to be more limited in scale. Consequently, rich insights into the human-environment interface have usually been limited to studies of local or regional practices and knowledge. This is both a weakness and a strength. On the one hand, the human sciences and humanities have not been able to create a global model of human-environmental interfaces. On the other hand, they have been able to demonstrate the complicated motivations of individuals and groups in shaping ecosystems.

Despite calls from scholars for interdisciplinarity, there remains a huge disconnect between environmental scholarship across the disciplines. Numerous scholars have recognized this fact, and recently, an editorial in *Nature* made the argument for the importance of the social sciences and humanities: "If you want science to deliver for society, you need to support a capacity to understand that society" ("Time for the Social Sciences" 2014). In 2012, *RESCUE,* a report commissioned by the European Science Foundation, Strasbourg, and European Cooperation in Science and Technology, Brussels, lamented the lack of interdisciplinary research collaborations and articulated the need for conceptual and methodological disciplinary integration from the earliest stages of new research projects (Jäger et al. 2012). As recognized in the *RESCUE* report, the social sciences and humanities have typically been auxiliary to the core agendas of scientific environmental research despite the fact that the environmental social sciences and humanities have been around for decades. For their part, the social sciences have been easier to integrate into scientific research. After all, human population patterns, economies, and governance frameworks are measurable and quantifiable. Likewise, historical and archaeological research has provided quantitative and qualitative data on environmental phenomena for developing and testing scientific theses (Carey 2012). Among the organizations that continue to play important roles in integrating the social environmental sciences are Future Earth, which absorbed the International Human Dimensions Programme on Global Environmental Change in 2014, and the International Social Science Council, which works closely with Future Earth.

On the other hand, ethnography, social and cultural history, environmental ethics, and postcolonial literary criticism have been tangential to environmental science. There are, however, several projects attempting to model an approach to bridging the humanities and sciences. Taking the lead in these is IHOPE, Integrated History and Future of People on Earth, originally a project of the International Geosphere and Biosphere Programme (IGBP). Since IHOPE was established in

2003, scholars involved in the project have consistently articulated the position that social scientists and humanists need to be more fully involved in earth system studies (Hornborg and Crumley 2006; Mosley 2006; Costanza, Graumlich, and Steffen 2007; Costanza et al. 2012; Sörlin 2012; Davies and M'Mbogori 2013). In 2013, Uppsala University in Sweden created a formal center for IHOPE. More recently, UNESCO's International Hydrological Program has commissioned a series of studies on water that promise to integrate a broader range of disciplinary approaches (Hassan 2011).

Joining in the spirit of these projects, RoA is unique in three ways. First, it focuses specifically on global river systems in the Anthropocene. Addressing rivers is not meant to isolate rivers from lakes, aquifers, coastal waters, sewage infrastructures—or even the complex web of flora and fauna that rely on them. Rather rivers serve as a practical frame within which to center research as well as a useful locus for analyzing flows, intersections, and cycles that are central to understanding the human-environment nexus. Second, the RoA Network integrates representatives from academia, government, and nongovernmental organizations who represent the research, policy, education, and community sectors. Third, public scholarship and community practice are central to the mission of the project and the crafting of the RoA Framework. Unlike many transdisciplinary projects, practicing artists and representatives from community organizations have been part of RoA from the beginning—contributing to the questions we ask and the methods we pursue. The essays in this volume represent only a fraction of the work being done by members of the RoA Network, which involves traditional research on the environment as well as art exhibitions and oral history projects that examine the relationship between communities and their waterways.

．．．

This volume is the first of several planned edited volumes focused on interdisciplinary and transdisciplinary approaches to the Anthropocene. By putting disciplines in dialogue with each other, it seeks to begin the path toward a more transdisciplinary engagement. The book's introductory chapter, "Anthropocenes: A Fractured Picture," offers a brief history of the "Anthropocene" both as a historical concept and as an empirically measurable phenomenon. It suggests that scholars should reject any easy notions about what defines the Anthropocene. Instead, they should embrace its complexities and inconsistencies. Doing so as part of a larger effort to pursue transdisciplinary research and policy will help us create more robust solutions to the problems facing humanity in the twenty-first century. The chapters that follow exist in dialogue with the introduction and are divided into three parts representing the many ways that scholars construct research questions, frame problems, and define methodologies.

Part 1, "Methods," demonstrates multiple modalities for interdisciplinary research, policy, and community-based environmental work. Andy Large, David Gilvear, and Eleanor Starkey's contribution to this volume focuses on the gaps in knowledge between assumptions and assessment in riverine research and policy. The authors argue that international and national ecosystem frameworks have often been hampered by a lack of data-based evidence on socioenvironmental entanglements in Anthropocene riverscapes. They propose an ecosystem service approach that uses citizen science to create structured catchment condition assessments, providing the necessary quantitative data necessary for making better policy decisions.

Turning to Africa, Sina Marx examines the political ecology of irrigation management in the Blue Nile Basin. Her analysis looks at institutions and discourses surrounding the Koga project, a large-scale irrigation scheme begun in the 1970s. Her work articulates the complicated politics of transboundary water management, particularly in an age of climate variability. She shows that the Koga project's success is complicated by the shifting contexts and desires of international agencies, governments, local leaders, and local publics. She demonstrates that a multiscalar analysis of these institutions, using techniques derived from social sciences such as anthropology, is essential to addressing water and food security challenges.

Moving the discussion of the Anthropocene further into the realm of the humanities is Celia Deane Drummond's critique of the Anthropocene's narrative. Focusing on the ethical implications of the concept, she suggests that the Anthropocene, as an apocalyptic narrative, imposes limits on how we conceive of our future in moral and ethical terms. This, she says, promotes a tendency to write the story of the environment in sweeping generalizations. Noting the dangers of fatalism in the grand narrative of the Anthropocene, she argues for scholars to focus on the "*local river system* and its specific instances of human/natural interactions." Doing so will help foster a "version of postnatural politics" that emphasizes the capacity to shape the future in tandem with other natural systems.

Concluding this section is a piece by Kenneth S. Lubinski and Martin Thoms that presents a sequence of challenges to scholars who wish to pursue transdisciplinary water research. The authors argue for the importance of defining measures of success. They emphasize the fact that while scholars have played (and continue to play) important roles in mediating between research, education, and policy, there is a potential dissonance between their goals and the conservative tendencies of sociopolitical institutions. In its attempt to establish a baseline from which transdisciplinary river research can move forward, this thought-provoking essay provides a clear framework for future water research.

Part 2, "Histories," examines the ways that our histories and research agendas shape water research. Jan Zalasiewicz, Mark Williams, and Dinah Smith connect the deep history of anthropogenic change to the rapidly changing conditions of the Anthropocene. Tracing changes in human activity and its effect on the geology of the English fenland, they show the long imprint of humanity on the environment. In so doing, their work suggests that reshaping the environment is not the same as effecting global changes in the environment. In this sense, their work underlines their broader research project on the Anthropocene, which argues that the Anthropocene might best be dated—at least in a geological sense—after 1945.

Michel Meybeck and Laurence Lestel return the conversation to European water management systems in their study of the Seine since 1880. Echoing Large, Gilvear, and Starkey's essay, Meybeck and Lestel show how cities affect water quality along the length of a river system. A city can have both upstream impacts through activities such as damming or timber rafting and downstream impacts through nutrients and toxic material inputs. Using long-term historical data on the Seine that they have collected over the past twenty-five years, they show that Paris and its river is a perfect exemplar of an Anthropocene river system. Their research is a model for how to analyze freshwater systems as dynamic historical entanglements of human and natural systems.

Philip Scarpino's essay is a history of the concept of the Anthropocene—specifically, from the perspective of an environmental historian. In tracing the long history of the idea, he weaves it together with the history of environmentalism in the twentieth century, arguing that it was the culmination of a series of ideas that developed over decades. He continues by making the point that scholars need to be careful when using the concept of the Anthropocene as a heuristic tool. Culture, he argues, is historically contingent and manifests itself in different ways in different contexts. As such, any study of entangled natural systems and human systems must take into account variable local conditions and not assume culture is a "single, undifferentiated variable."

In the final section of the volume, part 3, "Experiences," the authors explore the multiple ways that individuals and communities are shaped and reshaped by their interactions with their environments. The first essay, by Helen Berry, is a history of the Great Tyne Flood of 1771, which took place in Newcastle. Her study gives us insight into a city on the verge of industrialization at the dawn of the Anthropocene. It explains how Northeast England responded to one of the most catastrophic natural disasters that it has ever faced, posing questions about how states, municipalities, and community organizations respond to crises. Berry encourages us to think about the role that historical storytelling plays in shaping attitudes about our environments and societies, both past and present.

Stephanie C. Kane moves the discussion to Singapore, exploring the challenges of island nations in the age of the Anthropocene. Looking at the urban infrastructure of Singapore—its dams and drainage systems built both for flood control and to provide freshwater to the population—she demonstrates the complex dynamics between cultural, geological, and technological structures. She argues that a key feature of the Anthropocene is a state of never-ending tension between humanity's attempts to control and predict and the irrepressible power of geophysical systems. Inherent in these tensions is a form of coevolution, one in which society, technology, and environment are constantly reshaping each other, all the while transforming cultural assumptions and ways of knowing.

Mary Miss and Tim Carter present a case study that demonstrates the power of transdisciplinary collaboration. It reports on the first part of City as a Living Laboratory (CALL), a multiphase project in Indianapolis. This collaboration brings together scientists and artists—as well as government agencies, including the USGS—to address the issue of education and civic understanding of waterways. Through a series of installations based throughout Indianapolis, the project has focused on getting citizens to recognize the profound importance of local waterways to their lives. It underlines one of the central premises of the RoA project: addressing the challenges of the Anthropocene necessitates a wholesale cultural transformation in attitudes, expectations, and relationships to river systems. CALL shows one way that transdisciplinary collaboration can help effect this change.

In the final essay of the volume, "What Is a River? The Chicago River as Hyperobject," Matt Edgeworth and Jeff Benjamin use a phenomenological approach to examine the massive transformation of the Chicago River. For the past two centuries, humans have reworked its flows to such an extent that it has become a "hyperobject"—a concept developed by Tim Morton. Edgeworth describes the Chicago River as a thing that has become "large and multifaceted and spread out through time [with facets that are] hidden and inaccessible, phasing in and out of human awareness" (Morton 2013).

. . .

The contributions to this volume reveal that there is great value in interdisciplinary approaches that appreciate and explore the tensions inherent in different forms of research and practice. They suggest that a scholarly consensus on questions, methodologies, answers, and outcomes might not be as important to the success of interdisciplinary or transdisciplinary projects as the participants' willingness to allow space for ambiguity. Nevertheless, there are a number of common themes that emerge over the course of the volume.

PROBLEMS OF SCALE

Choosing different temporal and geographic scales creates different research and methodological problems. Long time frames (e.g., the millennia encompassed by Zalasiewicz, Williams, and Smith's work on English fenlands) show how profound human-nature interactions can be over the longue durée. However, the role of humans as individual actors (e.g., those in Berry's chapter on the Tyne flood of

1771) can get effaced by the long sweep of history. Likewise, while important to our understanding of global processes, the geographically sweeping analyses so often found in work on earth systems are more useful at identifying challenges than providing guidance at the regional or local levels, which might require unique technological understanding, understanding of sociopolitical structures, and cultural acuity and local knowledge (see Kane's chapter on the River Valley Planning Area in Singapore).

PROBLEMS OF INEQUALITY

The experience of the Anthropocene is hardly universal. Different regions can experience radically different Anthropocenes. The political ecology of the Blue Nile Basin described in Marx's chapter shows groups confronting problems fundamentally different from those experienced by Parisians over the past 150 years, as in Meybeck and Lestel's piece on the Seine. Likewise, our understanding of the environment and our expectations about our relationship to it are constructed through sociocultural structures—a theme central to Deane Drummond's essay on the ethics of the Anthropocene.

PROBLEMS OF AGENCY

Closely related to issues of scale and subjectivity is the notion of agency. Humans act as agents at multiple scales. As individuals, we make choices, and in this sense, we might be seen as rational agents in transforming our environments—primarily through our consumption patterns. Our choices have direct, observable consequences. Yet, while individuals have the capacity to consciously effect change, our actions are limited by the contexts in which we find ourselves. Each of us is shaped by our material, sociopolitical, and cultural worlds. Marx's truism holds for our understanding of individuals' relationships to their environments: "Men make their own history, but they do not make it as they please; they do not make it under self-selected circumstances, but under circumstances existing already, given and transmitted from the past." Furthermore, even if, as individuals or groups, we could be rational actors, making rational choices all of the time, anthropogenic environmental consequences are an emergent property of human systems—social, political, cultural, economic, and so on. Even our rational choices—either individually or collectively—can lead to unintended consequences. How we understand human agency and how we predict impacts shape research agendas—and consequently how we respond to the challenges of the Anthropocene.

In the end, this book does not prescribe a method for approaching these problems. Rather, it demonstrates the value of putting our disciplines into dialogue with each other. This book's chapters, full of rich case studies and thoughtful analysis, suggest the potential inherent in a research environment "without stable boundaries between the disciplines."

NOTES

1. "Enfin, à l'étape des relations interdisciplinaires, on peut espérer voir succéder une étape supérieure qui serait << transdisciplinaire >>, qui ne se contenterait pas d'atteindre des interactions ou réciprocités entre recherches spécialisées, mais situerait ces liaisons à l'intérieur d'un système total sans frontières stables entre les disciplines."

2. Much of the material in the section is borrowed from Kelly 2014.

3. While many scholars have focused on dating the Anthropocene to a period within the past five hundred years, there is a body of scholarship that argues for a "deep history" of the Anthropocene going back thousands or tens of thousands of years. See Ruddiman 2003, 2007, 2013; Braje and Erlandson 2013a; Braje and Erlandson 2013b; Smith and Zeder 2013; Barnosky 2014; Lewis and Maslin 2015; Zalasiewicz 2015.

ACKNOWLEDGMENTS

The volume editors would like to thank the following individuals for their support and advice during the conference and the preparation of this volume: Fredrik Albritton Jonson, Oscar Aldred, Rebecca Allan, Scott Ashley, Simon Atkinson, Anik Bhaduri, William Blomquist, Stephen Bridges, Kristen Cooper, Owen Dwyer, Tom Evans, Gabe Filippelli, Alex Hale, Kate Harris, Tom Iseley, James Jewitt, Vicky Keramida, Mark Kesling, Kathy Lamb Kozenski, David Lewis, Hines Mabika, Pam Martin, Fiona McDonald, Scott Morlock, Caron Newman, Claudia Pahl-Wostl, Kristi Palmer, Nasser Paydar, Jim Poyser, Martin Risch, Silvia Secchi, Nigel Thornton, Molly Trueblood, Sam Turner, Kody Varahramyan, Charles Vörösmarty, Chance Wagner, and Bill Werkheiser. We would especially like to thank James Yoder for his assistance in compiling the bibliography and editing.

Institutional support for the first Rivers of the Anthropocene Conference was provided by the Indiana University Office of the Vice President for Research's New Frontiers New Currents Grant, Newcastle University, the IUPUI Arts and Humanities Institute, Keramida Inc., the IUPUI Library, and the British Consulate-General Chicago. Our partners included the Center for Earth and Environmental Science at IUPUI, the Center for Urban Ecology at Butler University, the Center for Urban Health at IUPUI, Reconnecting to Our Waterways, the daVinci Pursuit, the Newcastle Institute for Research on Sustainability, and the Geography Educators' Network of Indiana.

Anthropocenes

A Fractured Picture

Jason M. Kelly

"There was no such thing as the Scientific Revolution, and this is a book about it" (Shapin 1996, 1). So began Stephen Shapin's *The Scientific Revolution,* a work, concise and smart, that embodied an approach to the history of science termed "the social construction of science." Shapin argued that if we are going to talk about a "scientific revolution," then we need to see it not simply as a historical event, but as a product of trends in twentieth-century historical writing. Following a pattern laid down as early as the eighteenth century, much twentieth-century writing conceptualized the Scientific Revolution as the linear unfolding of reason—a process in which discovery built on discovery, inevitably ushering in the modern world. The Scientific Revolution, in this story, completely transformed the intellectual landscape and allowed people to imagine natural phenomena in fundamentally new ways. However, as Shapin countered, if there was a Scientific Revolution, it was not a single moment but a set of processes that took place over hundreds of years and unfolded unevenly across different fields of study. The changes in understanding and practices that did take place were initially limited to a relatively small group in society, and these people needed to legitimate their claims within dominant intellectual and social frameworks. In fact, what they could claim as knowledge was hotly contested both within their various scientific communities and beyond. The Scientific Revolution was a powerful way for thinking about changes in early modern science, but it was neither so linear, complete, nor isolated from sociocultural concerns as moderns had been tempted to imagine.

What Shapin was arguing was hardly iconoclastic when he wrote in 1996.[1] In fact, his book was the product of decades of research that overturned triumphalist accounts of the history of science (Feyerabend 1975; Bloor 1976; Latour and

Woolgar 1986; Shapin and Schaffer 1986; Haraway 1988; Latour 1988; Daston and Galison 1992; Shapin 1995; Cetina 1999; Daston and Galison 2010). This scholarship suggested that science was neither internally rational and objective nor removed from its historical context. Science was a sociocultural practice like any other. At its most general level, this approach to the history of science—sometimes referred to as scientific constructivism—asks the question, how does something become deemed "true" or "false" in science?[2] How are decisions made, problems constructed, experiments formulated, solutions articulated? Shapin and his scientific constructivist colleagues argue that no scientific knowledge exists in a vacuum; the questions scientists ask, the methods they use, the claims they make are in fact social constructions. Consequently, science is a social practice always mediated by culture, social structures, economics, politics, and religion, which shape its production and consumption in the laboratory and beyond. Importantly, their analyses are not necessarily focused on the validity of truth claims but rather on the forces that drive the search for truths, determine interpretations, or influence reception.

Shapin's and his colleagues' critique of triumphalist accounts of the Scientific Revolution is a useful framework for thinking about the so-called Age of the Anthropocene. As with "the Scientific Revolution," a term first used in the early twentieth century, "the Anthropocene" is a neologism, used widely only since the early twenty-first century (Crutzen and Stoermer 2000; Meybeck 2001; Steffen et al. 2004; Syvitski et al. 2005; Costanza, Graumlich, and Steffen 2007; Robin and Steffen 2007; Zalasiewicz et al. 2008; Chakrabarty 2009; Rockström et al. 2009; Armesto et al. 2010; Davis 2011; Steffen, Persson, et al. 2011; Zalasiewicz et al. 2011; Dibley 2012; Crutzen and Steffen 2016). The origins of both concepts can be traced back two hundred years before their wide use—to the Enlightenment in the case of the Scientific Revolution and to the middle of the nineteenth century in the case of the Anthropocene. As new concepts they had imaginative force, reflecting changes in contemporary attitudes about the past as well as a sense that the present was experiencing a revolution. It is not a coincidence that the term "Scientific Revolution" was adopted widely at a moment when relativity, quantum physics, logical positivism, and even psychiatry suggested major leaps forward in knowledge about the universe and human cognition. Likewise, it is not a coincidence that "Anthropocene" entered the popular lexicon at a crucial moment in our understanding of earth systems science, neurobiology, exoplanets, and wide-scale threats to the planet's ecosystems.

This essay examines the origins of the concept of the Anthropocene by comparing and contrasting nineteenth- and twenty-first-century attitudes to irreversible anthropogenic impacts on the earth. Doing so helps elucidate how our understandings of anthropogenic environmental transformation have been (and remain) entangled with the historical legacy of our social, political, and cultural worlds. It suggests that contemporary discussions of the Anthropocene have close historical connections to nineteenth-century thought, which was not value neutral and which often served the interests of European and American imperial powers. Because of this, this essay suggests that there is no such thing as a singular Anthropocene—like the Scientific Revolution, the category is embedded in wider sociocultural frameworks—and that it would be productive for scientists, humanists, policy makers, and others to engage with it in more nuanced ways. Fracturing the Anthropocene into Anthropocene*s* helps combat a tendency to oversimplify complex, historically emergent biophysical and sociocultural entanglements. In sum, there is no such thing as the Anthropocene—at least as we typically discuss it—and this is an essay about it.

. . .

In Europe, humanity's relationship with the earth changed dramatically in the nineteenth century. In just a few decades, a planet that had long seemed young became millions, then billions of years old. Its face, once etched and cracked by a single great flood, was now marked by eons of watery flows, fiery magmatic expulsions, and layers upon layers of briny sediments. Fossils, from microscopic plankton to gargantuan reptiles, indicated worlds that had come and gone. The biosphere, once imagined to be constant and unchanging, was in fact a world of constant flux. Plants and animals—even human beings—were no longer the fixed creations of an omnipotent and beneficent heavenly creator. Every creature was subject to change, development—even extinction—as internal mutations and ever-morphing environments altered the balance between resources and reproduction. The Renaissance's Great Chain of Being, which suggested an orderly and hierarchical relationship between the divine and the earthly, was broken. For increasing numbers of people, the new cosmology made a supreme being seem unnecessary and irrelevant.

Grappling with the work of people such as Hutton, Cuvier, Lyell, Wallace, and Darwin—with concepts of deep time, a planet with many geological ages, and a constantly changing natural world—necessitated that scientists and philosophers alike shed many of the last trappings of medieval Aristotelianism, Platonism, and Renaissance notions of providence and order. It forced them to resituate humankind in the grand order of natural processes. If Copernicanism had decentered earth's place in the universe, the revolutions of the early nineteenth century removed humans from the center of earth's history. In fact, the notion of deep time suggested that humans were relatively tangential to the course of natural history. Only a belief in the invisible hand of providence—of a deity that controlled the seemingly random processes of evolution—could promise a master plan in which the existence of humans was more than mere chance.

Even as contemporaries began to grapple with these facts, integrating them into their scientific models, philosophical categories, and historical narratives, many began to notice that humans seemed to be quickening the pace of environmental change. Taking the long view of the history of civilization, Humphry Davy argued in 1830 that humanity had initiated its own geological age.

> Were the surface of the earth now to be carried down into the depths of the ocean, or were some great revolution of the waters to cover the existing land, and it was again to be elevated by fire, covered with consolidated depositions of sand or mud, how entirely different would it be in character from any of the secondary strata; its great features would undoubtedly be works of man, hewn stones and statues of bronze and marble, and tools of iron, and human remains would be more common than those of animals on the greatest part of the surface. The columns of Pæstum, or of Agrigentum or the immense iron and granite bridges of the Thames, would offer a striking contrast to the bones of the crocodiles or sauri in the older rocks, or even to those of the mammoth or elephas primogenius in the diluvial strata. And, whoever dwells upon this subject must be convinced, that the present order of things and the comparatively recent existence of man, as the master of the globe, is as certain as the destruction of a former and different order and the extinction of a number of living forms which have now no types in being; and which have left their remains wonderful monuments of the revolutions of nature. (Davy 1830, 146–47)

Writing in 1848, the president of the Ashmolean Society, Hugh Edwin Strickland, observed that humans were becoming prime movers in the extinction of species.

> It appears, indeed, highly probably that Death is a law of Nature in the Species as well as in the Individual; but this internal tendency to extinction is in both cases liable to be anticipated by violent or accidental causes. Numerous external agents have affected the distribution of organic life at various periods, and one of these has operated exclusively during the existing epoch, viz. the agency of Man, an influence peculiar in its effects, and which is made known to us by testimony as well as by inference. (Strickland 1848, iii)

The planet's deep history was entering a new phase. The human population was booming. The consumption of resources was increasing. With this came a concomitant effect on natural systems.

In the 1830s, Charles Lyell, the geologist so influential on Charles Darwin, described the destructive tendencies of humankind in the second volume of *Principles of Geology* (1832). Human migrations, he argued, were responsible for introducing foreign species that devastated local ecologies. One hundred fifty years before Alfred Crosby, he described a version of the "Columbian Exchange" in which Old World horses, cattle, and hogs upended and displaced American species (Crosby 1973). Lyell questioned the ultimate benefits of draining fens and clearing forests. Dubious about anthropocentric models of progress, he mused, "It admits of reasonable doubt whether, upon the whole, we fertilize or impoverish the lands which we occupy"(Lyell 1832, 2:146–47). In sum, he argued, "Man is, in truth, continually striving to diminish the natural diversity in the *stations* of animals and plants in every country, and to reduce them all to a small number fitted for species of economical use. He may succeed perfectly in attaining his object, even though the vegetation be comparatively meagre, and the total amount of animal life be greatly lessened" (147–48; original emphasis).

Critics would soon term the man whom Lyell had in mind *homo oeconomicus,* a pejorative neologism used to connote a modern person ruled by rationality, markets, and selfish individualism. *Homo oeconomicus* could be found perusing his mills in Manchester or planning new mineshafts for his holdings in Durham. Economic man saw copses, meadows, and fens as wastes waiting to be turned into productive cropland or factory floors.[3] He saw European imperialism as—if not good—a necessary evil that would benefit both the conqueror and the conquered. Imperial commerce, industrialization, and urbanization would bring wealth to the metropole while imposing European religion, morals, and education on inferior peoples. Reshaping global ecologies, imperialism would improve foreign lands along European models by intensively extracting natural resources and cultivating cash crops. The governor-general in India, Charles John Canning, 1st Earl Canning, reflected this attitude when he stated in December 1858:

> As regards the sale of waste lands [in Awadh], there can be no question of the substantial benefits, both to India and to England, which must follow the establishment of settlers who will introduce profitable and judicious cultivation into districts hitherto unclaimed. His Excellency in Council looks for the best results to the people of India, wherever in such districts European settlers may find a climate in which they can live and occupy themselves without detriment to their health, and whence they may direct such improvements as European capital, skill, and enterprise can effect in the agriculture, communications, and commerce of the surrounding country. He confidently expects that harmony of interests between permanent European settlers and half civilized tribes by whom most of these waste districts or the country adjoining them are thinly peopled will conduce to the material and moral improvement of large classes of the Queen's Indian subjects. (*Papers Relating to Land Tenures and Revenue Settlement in Oude* 1865, 251–52)

Through conquest, expropriation, settlement, commerce, and technology, *homo oeconomicus* attempted to bend the planet and its peoples to the desires and ideologies of the European and American bourgeoisie and their empires.

· · ·

There was little doubt in the mind of learned contemporaries that even though the planet had been constantly in flux over the course of its history, something unprecedented was taking place: humans seemed to be having an increasingly outsized (and devastating) impact on their environments. Some voiced concerns about humanity's attempts to control natural processes. In *Frankenstein* (1818), for example, the consequences were tragic. In pretending to be like a god and attempting to master nature, Victor Frankenstein finds himself mutant nature's slave, his monster declaring, "You are my creator, but I am your master;—obey!" By the end of the novel, Frankenstein, psychologically broken, finally admits, "Man . . . how ignorant art thou in thy pride of wisdom!" It was a moral fable that resonated with many contemporaries and set a precedent for subsequent works, most famously *The Island of Dr. Moreau* (1896).

The adulteration of nature might open a Pandora's box of uncontrollable hybrids and monsters—a world of unintended consequences for humanity's hubris. Of course, these were intuitions and conjectures. There was no way that contemporaries could have known the extent to which they were transforming the planet. However, there were indications. A small but growing number of prominent examples, such as the dodo of Mauritius or the bison herds of North America, suggested that humans could wipe entire species from the face of the earth. Human industries, sewer systems, and habitation could dramatically transform water systems as well. Industrial waterways had become so polluted that by 1867 the water from the River Dee (Afon Dyfrdwy) near Chester was "so poisoned that, mixed with five hundred times its quantity of wholesome water, it was so deadly that no fish could live in it" ("The Salmon Fisheries Conference [Horticultural Gardens, South Kensington, 7th June 1867]" 1867, 155). By clear-cutting forests, contemporaries recognized that they could change the climate, though, to be clear, this was not always considered problematic. As Andrew Ure reported in 1831, "The improvement that is continually taking place in the climate of America, proves, that the power of man extends to phenomena, which, from the magnitude and variety of their causes, seemed entirely beyond his controul"(Ure 1831, 335).

In a sense, concern (or pride) over science's and technology's ability to manipulate nature—that is, recognizing that human-induced environmental changes could be permanent and measurable—was an antidote to the metaphysical displacement of humanity from the center of natural history. Emphasizing human agency in effecting environmental change and its responsibility for mitigating negative consequences helped reassert humanity's place in the natural world. It is not surprising, therefore, that the conservation movements of the nineteenth century reasserted (in secular terms) one of the major precepts of Christian theology: humankind's dominion over the earth.

Those alarmed about irreversible environmental change included George Perkins Marsh, who was, with people such as Frederick Law Olmsted, among the early and vocal advocates for conservation and the creation of nature preserves. In 1864, Marsh wrote *Man and Nature,* one of the period's most perceptive and influential warnings about anthropogenic environmental change: "The earth is fast becoming an unfit home for its noblest inhabitant, and another era of equal human crime and human improvidence, and of like duration with that through which traces of that crime and improvidence extend, would reduce it to such a condition of impoverished productiveness, of shattered surface, of climatic excess, as to threaten the depravation, barbarism, and perhaps even the extinction of the species"(Marsh 1864, 44). The conservation movement that he helped spur in the United States eventually resulted in a system of national parks, forests, and animal preserves.

The early conservation movement was, however, a product of its place and time. While criticizing the worst abuses of *homo oeconomicus,* conservationists were tied nevertheless to the structures of capital and empire. In fact, much of the impetus for conservation came from those who didn't want to waste natural resources, seeing them as economic and imperial tools—*national* resources that required state management and protection. In India, for example, Alexander Gibson, Dietrich Brandis, John McClelland, and Hugh Cleghorn called for the establishment of a forest service in response to deforestation caused by logging for an expanding railroad system and the navy (Das 2005; Beinart and Hughes 2007; Grove 1996). The railways, like so much colonial infrastructure, existed primarily for the extraction of Indian resources, which, as with the case of cotton, could also be both environmentally and economically devastating to the colony. However, in most publications of the period, it was not the machine of empire but rather the indigenous peoples and rapacious, immoral merchants who were blamed for the damage. One representative report stated that "the natives" used teak "without afterthought for the future," fabricating wooden dishes "chopped out of the heart of a tree that would make the mainmast of a man-of-war, and the rest . . . left to rot" ("The Forests of Pegu" 1856, 253). His solution was to follow the advice of Cleghorn and McClelland and found a forest department.

Major state legislation came with the India Forest Acts of 1865 and 1878, which set aside forests for conservation, bringing an end to the most egregious practices of clear cutting. However, there was a more insidious side to these pieces of legislation. They established guidelines for the expropriation and seizure of land considered unused, unclaimed, or waste. And, mirroring the enclosure acts that had deprived Britons of their commons over the previous centuries, the colonial authorities immediately began seizing tens of thousands of square miles (Beinart and Hughes 2007, 117–18). By 1900, approximately 85,000 square miles of forests had been taken by the government—nearly the size of the province of Bengal (Gadgil and Guha 1993, 134).

In Africa, where Europeans had wreaked havoc for hundreds of years—murdering, pillaging, destabilizing governments, destroying infrastructure, and enslaving millions to feed their economies—they now arrived with advanced weapons, intent on extracting the continent's biological and mineral wealth. The arrival of more and more Europeans shifted the relationship between humans and the local fauna. Animals that had been hunted at more moderate levels were pushed to extinction as European markets demanded exotic furs and ivory. Imperialists transformed economies, and in large swaths of Africa they created first a boom, then a bust, in animal commodities.

Wildlife preserves served the interests of empire, protecting valuable commercial resources while providing elites with continued access to big game hunting. By the 1850s, the commercialization of African hunting led colonial administrators to establish preserves in the Knysna and Tsitsikamma forests, primarily to protect elephants (McCormick 1991, 9). By the 1890s, nearly all big game in British Africa fell under some form of administrative protection. A series of game laws adopted in the various colonies promised fines and jail time for those who hunted without permission. Nevertheless, governors still sold licenses to tourists who wished to hunt. One tourist guide from 1893 offered helpful hints to these hunter-tourists. In the (unlikely) event of being attacked by a lion, one should shoot it between the eyes or, failing that, in the shoulder, which would break its bones and prevent a "deadly spring." African elephants could not be shot between the eyes like Indian elephants, and hippopotami were to be shot beneath the eye and ear (*Brown's South Africa* 1893, 78–80).

Since big game exports could bring the colonies little revenue (by the 1890s, ivory exports had plummeted) hunting licenses provided a means for the state to squeeze just a bit more from its natural resources. And for European and American elites, this offered the thrill of an exotic hunt, which they could recount to their peers at private clubs in Paris, New York, Berlin, and London or in adventure narratives that were all the rage at the height of empire. When Theodore Roosevelt wrote *African Game Trails* about a hunt he took in 1909, he highlighted the dangers of the expedition: "During the last few decades, in Africa, hundreds of white hunters, and thousands of native hunters, have been killed or wounded by lions, buffaloes, elephants, and rhinos. All are dangerous game: each species has to its grewsome [*sic*] credit a long list of mighty hunters slain or disabled"(Roosevelt 1910, 72). At the end of the volume, he proudly listed his and his brother's kills in a table: 9 lions for Teddy, 8 for Kermit; 8 elephants for Teddy, 3 for Kermit. Together, they killed 512 animals (Roosevelt 1910, 532).

Unsurprisingly, the colonial administrators' efforts to control land and animals fell unevenly along class and racial lines (MacKenzie 1997, 201–60; Steinhart 2006). Some critics blamed the decimation of African species on indigenous groups—often with an explicit sense of moral and cultural superiority. François Coillard, for example, argued that it was "native hunters themselves who, totally destitute of conscience in this respect, are hastening the extermination of certain species"(Coillard 1897, 638; see also MacKenzie 1997, 233). The game regulations, which limited hunting over the last half of the nineteenth century, increasingly restricted traditional African hunting techniques in favor of guns, considered to be more humane (MacKenzie 1997, 209). Beginning in 1891 in Natal, for example, it was illegal not only to use "nets, springes, gins, traps, snares, or sticks" to catch animals and birds but also to own them with the intent to hunt (House of Commons 1906, 14). Firearms, however, were also regulated by colonial administrations, which sought to keep them out of the hands of African subjects. So, even as imperial governments argued for the use of guns in hunting, it also prohibited gun ownership for "any person of the native tribes of this Colony."[4] In other colonies, the government and various civic preservation societies used the cause of conservation to decide which groups should have the right to hunt. As John MacKenzie has noted, in Kenya, the Ndorobo would be encouraged to abandon hunting in favor of herding, but the Kamba, who were not reliant on hunting to subsist, would be banned from the activity (MacKenzie 1997, 215).

In effect, conservation legislation increasingly limited equal use of land and natural resources in favor of the European colonists. Walling off preserves from hunters without licenses was a nineteenth-century parallel to the aristocracy's claims to private hunting grounds in early modern Europe. And, as with the eighteenth-century Black Acts, which were used to control England's rural populace, imperialists in places such as the Cape Colony created legislation to wall off property over which they claimed control.

The close relationship between conservation, imperialism, and race is just one example of how easily environmental discourses and practices can become entangled with sociocultural, political, and economic fields. This observation aligns with what scientific constructivists argue: the history of science cannot be disentangled from broader cultural forces. This observation does not make the practice of conservation any less valid but it does reveal its potential to serve interests and ideologies beyond its stated objectives.

. . .

It is clear that by the early nineteenth century, what we might call an "anthropocenic consciousness" was emerging among the European and American scientific community. What I mean by this is that some people were beginning to recognize that humans were making potentially permanent changes to the earth and that this could be corroborated by empirical evidence. Developing this new understanding of humanity's relationship to the planet also necessitated policy changes in order to mitigate humanity's most devastating environmental effects. These adaptations, often favoring the interests of elites in Europe and America, included changes in forest practices, hunting, sewage infrastructure, and even consumption patterns.

Over the course of the next 150 years, the development of an anthropocenic consciousness was an uneven and protracted process. Despite the fact that early nineteenth-century writers articulated many of the basic concepts that scholars typically associate with the Anthropocene—despite evocative concepts such as a "silent spring" or "Gaia" in the twentieth century—it has been only in the first decade of the twenty-first century that scientists have coined a term that seems to have resonated with both the academic and general publics (Carson [1962] 2002; Lovelock 1974, 1983, 2000; Margulis 2008; Steffen, Grinevald, et al. 2011). There are now academic journals devoted to the Anthropocene, and new books on the topic are appearing at a rapid rate. Scholarly forums debate the definition of the term and the way that it might affect their disciplines. Academic discussions about "the Anthropocene" are beginning to spill over into popular culture, making the covers of *The Economist* and *National Geographic* and being reviewed in articles in the *New Yorker* and *The Guardian*.

With a conceptual lineage that goes back two hundred years, the Anthropocene brings with it a host of scientific, philosophical, and cultural accretions. This fact is never part of the popular discussion and is rarely examined in academic literature, except for the obligatory nod to Will Steffen, Jacques Grinevald, Paul Crutzen, and William McNeill's essay, "The Anthropocene: Conceptual and Historical Perspectives" (2011).[5] It is, however, implicit in some of the more nuanced scholarship. Take, for example, the Subcommission on Quaternary Stratigraphy's Working Group on the Anthropocene, chaired by Jan Zalasiewicz. Its task since 2009 has been to decide whether we can associate the Anthropocene with an identifiable and global "geological signal." If there were a "geological signal," then the International Union of Geological Sciences might vote to designate the Anthropocene a new geological age—symbolized by placing "golden spikes" (also known as a Global Boundary Stratotype Sections and Points) at representative points in the earth's stratigraphy. This working group has often been cited as key for asserting an "official" age of the Anthropocene, which, in August 2016, they put at 1945. In its mission documents, however, the working group notes the limits of its task: "the currently informal term 'Anthropocene' has already proven to be very useful to the global change research community and thus will continue to be used, but it remains to be determined whether formalisation within the Geological Time Scale would make it more useful or broaden its usefulness to other scientific communities, such as the geological community" (Subcommission on Quaternary Stratigraphy, International Commission on Stratigraphy 2015). In other words, its task is determined by the standards of the discipline of geology and that other research communities have used and will continue to use the term in different contexts. Implicitly, it recognizes that there is no single Anthropocene but rather multiple Anthropocenes that serve different but potentially complementary functions.

Thinking about the Anthropocene as a cluster of mutually complementary approaches recognizes the historically complicated origins of the idea and opens up the possibility of rich multidisciplinary dialogues that have the potential to reshape research and policy agendas. Likewise, it makes it easier to acknowledge the fact that while anthropocenic indicators—climate change, ocean level rise, water pollution, cultural awareness of environmental change, and more—may be globally measurable phenomena, they are not experienced equally around the planet. Further, this approach is more in line with actual usage in that the Anthropocene is a concept that serves a variety of sometimes incommensurable agendas and perspectives.

For example, an approach to the Anthropocene that focuses on geological indicators might find that while there were increased anthropogenic effects since the eighteenth century, a marked sedimentary divergence occurs only in the middle of the twentieth century, with the introduction of radioactive isotopes created by nuclear fallout—a nuclear Anthropocene. Moving away from geological indicators to mediums such as ice cores, tree rings, and coral and to other isotopic signatures such as $\delta^{13}C$ or $\delta^{15}N$, we might find Anthropocenes manifesting at different rates, times, and places (Dean, Leng, and Mackay 2014). Likewise, changes in biodiversity and a so-called Sixth Extinction might indicate still other standards and moments for the onset of the Anthropocene (Braje and Erlandson 2013). In fact, depending on one's preferred data point, the Holocene-Anthropocene boundary might be as long ago as ten thousand years or as recently as fifty years (Steffen et al. 2005; Steffen, Crutzen, and McNeill 2007; Steffen, Grinevald, et al. 2011; Ellis, Fuller, et al. 2013; Ellis, Kaplan, et al. 2013; Ruddiman 2013).

No matter what the data, however, most studies assume that the Anthropocene is a measurable biophysical phenomenon, usually ignoring the fact that neither the choice of key data points nor the concept of an Anthropocene is value neutral. The Anthropocene nearly always serves as a metanarrative of modernity—a narrative in which energy- and resource-intensive industrialization and capitalism have been accompanied by population booms, increased flows of goods and peoples, the central role of nation-states, and demands for improvements in quality of life. It is a story in which humans have exploited the environment at unprecedented and ever-expanding rates, soon finding that their local actions have consequences on global scales (Kelly 2014). The Anthropocene becomes a category for critique—a way to define excess, limits, thresholds, and boundaries (Meadows et al. 1972; Chakrabarty 2009; Rockström et al. 2009; Dibley 2012). In effect, it becomes a standard by which to measure human action and hold societies accountable for their actions—an ethical framework. And, as scholars of environmental ethics, environmental justice, and ecocriticism suggest, this standard is often dominated by Eurocentric assumptions and interests (Egan 2002; Mosley 2006; Timmons Roberts 2007; Sze and London 2008; Ottinger and Cohen 2011).

Given the historical context in which anthropocenic consciousness emerged, it is not surprising that "the Anthropocene" is a term used in both descriptive and prescriptive senses. From its origins, the term "Anthropocene" was meant to convey an objective description of the world as well as to suggest a moral imperative to respond to the state of this world in appropriate e ways. In David Hume's sense, the Anthropocene serves dual functions as an "is" and an "ought." Take for example what might be considered the founding document of twenty-first-century research on the Anthropocene, Paul Crutzen and Stoermer's article "Anthropocene" in the International Geosphere-Biosphere Programme newsletter in 2000. After describing the conditions of anthropogenic change since the Industrial Revolution, they conclude:

> Mankind will remain a major geological force for many millennia, maybe millions of years, to come. To develop a world-wide accepted strategy leading to sustainability of ecosystems against human induced stresses will be one of the great future tasks of mankind, requiring intensive research efforts and wise application of the knowledge thus acquired in the noösphere, better known as knowledge or information society.

An exciting, but also difficult and daunting task lies ahead of the global research and engineering community to guide mankind towards global, sustainable, environmental management. (Crutzen and Stoermer 2000, 18)

In this short summary, Crutzen and Stoermer set the scientific agenda for research on the Anthropocene: to serve the needs of society by being embedded in the process of making public knowledge, guiding policy decisions, and advocating for proper environmental management policies. According to them, the task of the global research and engineering community is not simply descriptive, and it is certainly not value neutral. Deciding what ought to be done to solve a looming crisis for humanity is a fundamentally moral position that necessitates defining responsibilities and obligations as well as distinguishing between "good" and "bad" behaviors and responses.

. . .

Recognizing that the concept of the Anthropocene is already laden with a multiplicity of meanings, I would like to conclude by summarizing several ways that scholars might engage more productively with the concept across disciplines. These observations emerge out of the discussions of the RoA Working Group, which met in Indianapolis in January 2014.

First, given that human-induced environmental change continues to transform the planet in both predictable and unpredictable ways and given that researchers would like to create a framework for responding to these changes, it is not enough to simply understand biophysical environmental processes. It is necessary that researchers also understand the sociocultural processes that drive human-induced environmental change. This is necessary because things such as cultural beliefs can limit responses to ecological crises and therefore contribute to environmental feedback loops. Only through an understanding of the ways in which religious, economic, cultural, ideological, and political processes function—integrating them into our analyses of environmental processes and embedding them in our policies—are we likely to produce robust responses to the environmental challenges we face. This requires multidisciplinary teams of researchers, policy makers, and community stakeholders articulating agendas together and collaborating in the analysis of the human-environment interface.

Pursuing a multidisciplinary research agenda that integrates scientists, social scientists, humanists, artists, policy makers, and community stakeholders requires recognizing that "the Anthropocene" is a fluid signifier. The term encompasses a bundle of emerging concepts that reflect discipline-specific cultures, methodologies, and epistemologies. In other words, the Anthropocene is not a single thing, entity, or ideal; it is a category onto which different groups map multiple, and sometimes conflicting, ways of knowing and/or describing the world. Because of the many different ways that researchers approach these Anthropocenes—defining problems, asking questions, devising methods, and even articulating truth claims or uncertainties—collaborative work necessarily generates tensions among multidisciplinary participants. This can be incredibly productive when the project design focuses on articulating these tensions—from the very beginnings of the project discussing how different approaches frame or define boundaries differently.[6] Doing so reminds researchers that their disciplinary perspectives are subjective, historically situated, socially constructed models. It encourages participants to recognize that they tell only part of a larger story and that multidisciplinary cooperation may ultimately be the most effective way for all groups to achieve their ends.

Second, the Anthropocene is not simply an intellectual category for describing the environment. It is also a lived phenomenon that humans experience on a variety of scales. This simple fact can often get lost in discussions of CO_2 emissions and extinctions, or even in the critical analysis of the Anthropocene as an epistemological category. The emergent processes—the entanglement of environmental and sociocultural processes and structures—that characterize the Anthropocene have very real consequences for people's day-to-day lives. These consequences are experienced unevenly and therefore function quite differently in different contexts. Take, for example, the destruction of freshwater environments. The scale at which freshwater environments are threatened in the United States is surprising. Drought, overconsumption, industrial waste, agricultural runoff, and more mean that the state of over 40 percent of American waterways threatens aquatic life (U.S. Environmental Protection Agency 2015). In places such as the Great Lakes, the threat to aquatic life is near 100 percent (U.S. Environmental Protection Agency 2015). Nevertheless, these numbers do not take into account the United States' total threat to freshwater systems around the world. This is because from its position of safety, power, and wealth, the United States exports much of its pollution overseas. American consumption patterns and the international supply chain mean that much production for American markets is done in places such as Asia, where manufacturers pollute surface waters and drain aquifers. In fact, over 20 percent of the water footprint of the United States is beyond its shores (Mekonnen and Hoekstra 2011; Water Footprint Calculator 2015). The apparel industry alone accounts for significant water pollution in Asia, and dyeing textiles for American and European "fast fashion" has been particularly devastating (Brigden et al. 2012; Institute of Public and Environmental Affairs et al. 2012). Americans are free to consume without the worry of immediate consequences while pollution in foreign rivers is decimating species and increasing the incidence of death and disease.

The uneven experience of the Anthropocene is a product of its late eighteenth- and early nineteenth-century origins.[7] Even as Europeans and Americans were harvesting coal to replace the energy once provided by timber, water, and animals, they were also using their militaries to expand control over trading routes, territory, and natural resources. They extracted labor and materials while disrupting foreign political and economic systems. They imposed restrictions on trade and forced open markets to drive a sequence of industrial revolutions over the course of two hundred years. And, in so doing, they exploited, undermined, and under-developed foreign economies. As they absorbed the world's raw materials and pro-cessed them through coal-driven systems of manufacture, they caused their own economies to diverge from those in Asia, Africa, and Latin America (figs. 1.1, 1.2).

Economists and historians have discussed this "Great Divergence" in primar-ily economic terms, but there was an environmental side to the Great Divergence as well. European and American imperial capitalism did significant damage to the planet's ecosystems through CO_2 emissions, clear cutting, and industrial waste, and the consequences were experienced (and continue to be experienced) unevenly. By the last quarter of the twentieth century, the environmental move-ment forced reforms in Europe and the United States. This helped push some of the most costly environmental practices overseas, where poor and under-developed nations struggled to close the economic gap. Thus the Anthropocene is also a story of the unequal distribution of resources and environmental costs, amplified by political and economic structures and legacies. Consequently, any large-scale environmental solutions requires scientists to work with social scien-tists, humanists, policy makers, and local communities to both understand and design responses that address asymmetric power structures and the uneven con-sequences of global environmental change.[8] This move toward an environmental justice agenda is under way, but it requires environmental researchers to integrate into their work a deep analysis and critique of the structures of global capitalism as it relates to the human-environment nexus. Key topics include the following:

- the commodification of environmental resources (e.g., water) and knowledge (e.g., genomes or "improved seeds")
- the privatization of environmental commons (e.g., the Cochabamba Water War)
- the growth in the power, influence, and networks of nonstate actors, particularly through multinational corporations and the system of monopoly-finance capital
- the construction of poverty induced and sustained by systems of finance, trade, development, and technology
- planetary boundaries, cultural knowledge, and social practices at the local level
- displacement from intended and unintended human-induced environmental change (e.g., sea level rise, construction of dams)
- the distribution of environmental resources, risks, and responsibilities

Third, as a prescriptive category, the Anthropocene necessitates that research teams intentionally integrate questions from philosophical ethics and critical the-ory into their projects. After all, the ultimate purpose of nearly all anthropocenic research is to create standards for responsible policies and behaviors—effectively,

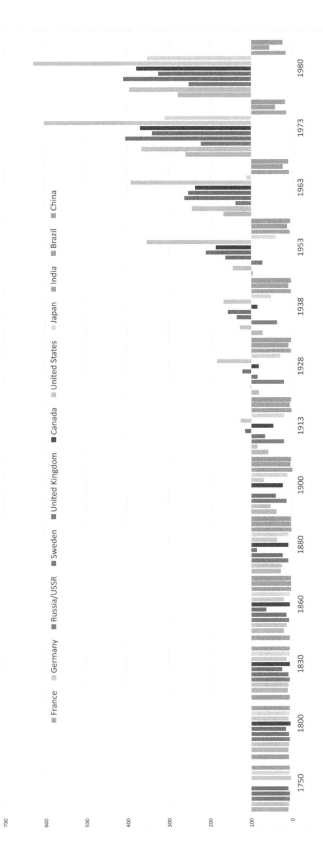

FIGURE 1.1. Per capita levels of industrialization (U.K. in 1900 = 100) derived from Bairoch 1982.

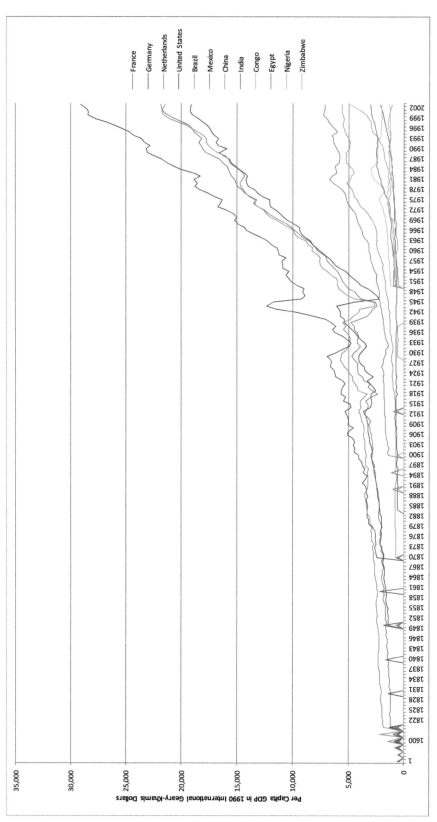

FIGURE 1.2. Per capita GDP based on Angus Madison's Historical Statistics for the World Economy: 1–2003 C.E., http://www.ggdc.net/.

guidelines for acting in the world, also known as normative ethics. And, since no moral code is objective and no ethical framework exists outside of historically situated sociocultural frameworks, the participation of specialists in morality, religion, history, art, and behavioral psychology is essential to the success of any project that seeks to most effectively address environmental problems—especially if the solution requires people to reimagine their cultural norms or transform their social practices.[9] Critical theory, on the other hand, is important in that its mission is to examine assumptions, discursive frameworks, and epistemologies that create or reproduce inequality. Critical theory provides ways for thinking about the role that research and policy agendas unintentionally play in perpetuating inequalities. And it points to new directions for addressing issues such as environmental justice.

The three observations outlined above have a common thread: multidisciplinarity. They suggest that research on the Anthropocene could better attain its ultimate goals—both descriptive and prescriptive—by building broader-based research and policy teams that integrate people from multiple (and sometimes epistemologically divergent) fields. When collaborative research projects are designed in such a way that participants can learn from one another, with the intent that only through pooling their specialties will they all be successful, the outcomes will be more fruitful. To do this, however, takes a sustained and intentional effort to integrate experts from across the disciplines. This is the ultimate goal of the RoA project—to provide an infrastructure and a set of standards for undertaking multidisciplinary research on the Anthropocene.

This essay began by arguing that there is no such thing as the Anthropocene and that this is an essay about it. What I want to suggest is that the Anthropocene (and what I term "anthropocenic consciousness") is not something that can simply be quantified, described, or measured. It is an emerging biophysical state as well as an emerging intellectual category. It is a thing both manifested in the physical world and manifested in our imaginations. As such, it is a fractured thing, or things—Anthropocenes. This realization can be very useful for researchers and can help us create more nuanced research and policy. Embracing this open-endedness can help us gain a clearer understanding of our assumptions, lead to more integrated cross-disciplinary engagement, and create better solutions to the greatest challenges facing humanity in the coming century.

NOTES

1. However, it should also be noted that Shapin published his book at the height of the "science wars" of the 1990s that pitted scientific realists (those who subscribe to a set of philosophical positions that claim science can reveal natural truth) against scientific constructivists. See Gross and Levitt 1994; Sokal and Bricmont 1999; Hacking 2000; Labinger and Collins 2001; Brown 2004.

2. For a summary of scientific constructivism and its historiography, see Golinski 2005.

3. See, e.g., Mill 1878, 324–25.

4. "All guns, or pistols, or gunpowder, found in the Colony in the possession of any person of

the native tribes of this Colony, or of any person of the native tribes of the countries adjacent thereto, without the written permission of the Governor as aforesaid, shall be seized and forfeited, whether the said gun or pistol be marked and registered or not; and the party in whose possession, as aforesaid, any such gun, or pistol, or gunpowder, may be found, shall be liable to a penalty not exceeding fifty pounds, or at the discretion of the Resident Magistrate to imprisonment for any term not exceeding two years" (Cadiz and Lyon 1891, 250).

5. See also Visconti 2014.

6. Projects that have done this effectively have recognized the benefits to their research outcomes (e.g., Dewulf et al. 2007; Mattor et al. 2013). Where project directors, participants, and institutions have integrated a dialogue about the ways that their respective disciplines frame discourse and create disciplinary boundaries, they have become more aware of their biases and thus more invested in the transdisciplinary process. From the beginning, these projects encourage "participatory modeling," an approach with analogues in other formats (e.g., "shared authority" in the field of public history) that allows participants to frame the problems and questions associated with the research. See Mollinga 2010.

7. The following paragraphs summarize the major outlines of a debate that continues to dominate discussions of modern world history. See Frank 1998; Landes 1999; Wong 2000; Vries 2001; Moore 2003; Duchesne 2004; Landes 2006; Allen 2009; Vries 2010; Parthasarathi 2011; Wallerstein 2011a, 2011b, 2011c, 2011d; Jonsson 2012.

8. While there is a prevalent subset of environmental research in the social sciences and the humanities that deals with environmental justice—and specifically environmental justice related to imperial and postimperial contexts—there remains a substantial disconnect between these discussions and more general discussions about the Anthropocene.

9. For an astute analysis of the problem of conflating biological description and normative ethics, see Thomas 2014.

PART ONE

Methods

Part 1 of this book explores various methodological approaches to the problem of the "Anthropocene" and, in so doing, challenges any simplified notion of what Anthropocene scholarship might look like. The concern here is first and foremost the implications of global anthropogenic environmental change, but it is also the ways that scholars, policy makers, NGOs, and communities might work together to respond to these challenges.

In different ways, the authors implicitly engage with methodological problems associated with scale. They are interested in how to take the abstract concept of the Anthropocene—the idea that it is an anthropogenic, historical, global phenomenon that has permanently altered the earth's systems (water cycles, climate, etc.) and has left a defined geological mark across the entire planet—and adapt it to regional and local conditions. They recognize the fact that scientific agendas, frameworks of governance, and even individuals' imaginations rarely operate at the global scale. Except for global modeling and high-level governance, such as the Paris Accords, most people's engagement with and understanding of the environment is much more localized. Even the Paris Agreement (United Nations Framework Convention on Climate Change), which was framed according to global principles, emerged out of localized interests and will be implemented differently throughout the world.

This disconnect between the abstraction of the Anthropocene and its lived realities is a challenge to researchers. For example, those working at the interface between science and governance recognize that biophysical systems rarely align with geopolitical boundaries. This is especially the case with rivers, which often flow across numerous geopolitical divides. Take the Colorado River. Reflecting on a visit to its delta in 1922, Aldo Leopold wrote of a vibrant ecosystem:

The still waters were of a deep emerald hue, colored by algae, I suppose, but no less green for all that. A verdant wall of mesquite and willow separated the channel from the thorny desert beyond. At each bend we saw egrets standing in the pools ahead, each white statue matched by its white reflection. Fleets of cormorants drove their black prows in quest of skittering mullets; avocets, wallets, and yellow-legs dozed one-legged on the bars; mallards, widgeons, and teal strand skyward in alarm. As the birds tool the air, they accumulated in a small cloud ahead, there to settle, or to break back to our rear. When a troop of egrets settled on a far green willow, they looked like a premature snowstorm. (Leopold 1968, 142)

That same year, individuals from states that intersected with the river signed the Colorado River Compact, a plan that set in process the decades-long siphoning of water from the river. The following years saw the construction of Lake Powell and Lake Mead. By the 1970s, coupled with the increased consumption of water upstream, the Colorado delta had shrunk until water no longer flowed to the sea, destroying a once healthy ecosystem and undermining the livelihoods of those who relied on its flow, including the Cucapá, who used the river for agriculture and fishing. A map of the Colorado River Basin published by the U.S. Department of the Interior, Bureau of Reclamation, in 2012 provides a metaphor for the challenges of working with transboundary river systems. In its summary map, "Colorado River Basin Water Supply and Demand," the basin itself stops at the U.S. border with Mexico (U.S. Department of the Interior, Bureau of Reclamation 2012). In effect, there is a historical mismatch between the geographical scales of river systems and the geopolitical scales of states.

Even when they do not cross international boundaries—even at the local level—sociopolitical frameworks shape how we respond to rivers. River governance is often a hodgepodge of overlapping public agencies, nonprofits, and private interests. One small stretch of a river in the eastern United States might be governed by federal agencies, such as the Army Corps of Engineers, the Environmental Protection Agency, or the United States Geological Survey; state departments of environmental management and natural resources; municipal authorities; local utilities; and citizens' groups. Conceptualizing the global nature of the Anthropocene in the context of regional or local affairs reveals the difficulty of scaling the concept. What does the Anthropocene mean to a local council or municipal government? How might it transform the decision-making process? In a democratic society, what are the implications for a disconnect between local conditions and global challenges in the minds of voters?

In chapter 2, Large, Gilvear, and Starkey ponder the problem of shifting baselines. Across large distances and swaths of time, capturing micro-level data to establish both site-specific and systemwide change is difficult. Their solution is to merge the framework of ecosystems services with open data and citizen science as a new method for capturing information. In chapter 3, Marx turns to issues of scale and power. In her words, humankind is "not a single global agent."

Both the causes and the consequences of environmental change are experienced unevenly. Looking at the Koga water projects in Ethiopia, she shows that even narratives about how to respond to environmental change operate differently at different scales. Drummond continues the theme of narrative by focusing on the story of the Anthropocene itself. She argues that all narratives—especially historical ones—embed ethical constructions. Her essay emphasizes the power of exploring these stories. In chapter 4, Lubinski and Thoms explore the relationship between scholars and society. Whereas Large, Galvear, and Starkey consider how to develop a methodology that involves citizens in the research endeavor, Lubinski and Thomas ask how scholars can keep issues relevant to the public and high on the priority list of policy makers. Their answer is that a key element of scholarly methodology is public engagement.

In sum, these chapters suggest that the Anthropocene—as both an intellectual concept and a lived experience—might encourage scholars to rethink the practices and assumptions built into their research practices and institutions. The geophysical-sociocultural shifts of the Anthropocene, new baselines and accelerated change, may require new modes of scholarship better suited to these new contexts.

2

Ecosystem Service-Based Approaches for Status Assessment of Anthropocene Riverscapes

Andy Large, David Gilvear, and Eleanor Starkey

RIVERS OF THE ANTHROPOCENE AND KEY DRIVERS OF GLOBAL CHANGE

Rivers are of immense importance, geologically, biologically, historically, and culturally, and they are central to many of the environmental issues that concern society (see, e.g., Sponseller, Heffernan, and Fisher 2013). It is clear, however, that we are entering an era in which humans are accelerating and decelerating natural processes and altering, creating, and destroying ecosystems at "an astonishing pace" (Syvitski 2012, 12). Gaffney (2009, 1) has described this as "moving out of the Holocene envelope," also highlighting the fact that in an "astonishingly short period" of 250 years, humans have developed the capacity to alter the global earth system in ways it has not been altered for millions of years. Pastore et al. (2010) highlight four principal drivers of hydrological change in river systems: water engineering, land cover change, climate change, and human decision making—all of which have provoked worldwide adjustments in terms of catchment-scale water stores and fluxes, biogeochemistry, and river morphology. Harrigan et al. (2013) demonstrate how multiple drivers, acting simultaneously but over differing time scales, drive stream-flow alteration. It is estimated that the annual, worldwide, deliberate shift of sediment equates to 57,000 million tonnes (Mt), an amount that exceeds that of transport by rivers from the land to the oceans (22,000 Mt) by almost a factor of three (Douglas and Lawson 2000; Price et al. 2011). In terms of the planet's river systems, we have clearly entered the "Anthropocene" (Crutzen and Stoermer 2000), when earth systems are becoming defined by human agency so profound that it is potentially affecting the stratigraphic record. Erosion and sedimentation offer a classic case of this process in action (Waters et al. 2016; Zalasiewicz et al. 2008).

The planet is now host to over seven billion people, and as of the first decade of the twenty-first[t] century, 50 percent of humanity was urbanized. Each and every one of us was born and lives in a river catchment; therefore, a number of key questions arise as to how we approach management of river systems, with their uneven range of pressures experienced under often intensely crowded conditions. Relatively natural or pristine segments of rivers are increasingly rare through-out much of the world. As Thorp, Thoms, and Delong (2006) attest, this makes it a formidable task to study and manage such systems in a human-dominated world. Ellis and Ramankutty (2008, cited in Schwägerl 2014, 38) make the claim that "only 22 percent of the earth's surface is still wilderness and only 11 percent of photosynthesis takes place in these wild areas." From this, they conclude that this new worldview of the biosphere constitutes a paradigm shift from it constituting "natural ecosystems with humans disturbing them" toward a vision of "human systems with natural ecosystems embedded within them" (Ellis and Ramankutty 2008, 445).

Yet after more than a century of research on rivers and their physical and biotic makeup, we still lack robust baselines as to how these freshwater ecosystems function. This paucity of reference points hinders widespread understanding of what ecosystem services are delivered by rivers either as natural systems with humans disturbing them or as human systems with remnants of natural aquatic ecosystems embedded in them. More and more, as we venture deeper into the new Anthropocene epoch (as defined by Waters et al. 2016), it is vital to gain this widespread understanding in forms that are accessible to scientists, planners, managers—and to the general public who live in these riverine landscapes or "riverscapes" (sensu stricto Wiens 2002). Scholars from a range of disciplines have traditionally framed problems of environmental change and degradation within disciplinary constructs; however, an increasingly important question is to what extent *trans*disciplinary perspectives on the relatively recently defined Anthropocene epoch can provide new ideas, new understanding, and better approaches to river management. Here, we define "transdisciplinary" as producing new frameworks of understanding and working that would not be achievable in individual disciplines alone or by using interdisciplinary approaches (where typically two disciplines come together to produce a more integrated approach).

In this chapter, we briefly explore what constitutes a "river of the Anthropocene" and introduce a methodology using free and ubiquitous software to assess river condition and status using physical (geomorphological) features as they relate to ecosystem service provision. The methodology is designed to have worldwide applicability, and we illustrate it here using the River Tyne, a medium-sized temperate river system in the United Kingdom. Enacting meaningful catchment or watershed-wide change in systems like the Tyne may appear a daunting task, but is perhaps more easily visualised as "seven billion collective decisions"—that is, envisaging a world where each and every one of us living in a particular watershed has a choice to make and a role to play. Using a subcatchment of the River Tyne, we therefore also briefly explore here the role of people and communities in "crowd-sourced knowledge partnerships" built through communities assessing and monitoring the "pulse" of their own watersheds. Such approaches, combining the rigor of small-scale studies with broader ecosystem-scale assessments (e.g., Nelson et al. 2009), as well as proper mapping, monitoring and assessment programmes (e.g., Naidoo et al. 2008; Langhans et al. 2013), are needed for more effective management of rivers in the Anthropocene through integration of new knowledge with changing societal goals.

THE CONCEPT OF DAMAGE

It is undeniable that today the vast majority of the planet's rivers are anything but "pristine" or even near-"natural" (see, e.g., Wohl 2013). Despite the fact that shoreline length and tributary junctions still provide key space in modern catchments for natural processes, worldwide we are no longer dealing with "natural" rivers. Here we use Newson and Large's (2006) definition of natural rivers as those requiring minimum management interventions to support system resilience and protect a diversity of physical habitat. While system resilience and habitat diversity are neither universal nor perpetual, their role increases with the proportion of the channel network within the fluvial system exhibiting a full interplay of unmanaged water and sediment fluxes with local boundary conditions. Catchment connectivity (and increasingly its lack) is therefore key. Over the past several decades, catchment management efforts worldwide have made major strides, but their overall effectiveness has typically been hampered by two things: lack of definition of what constitutes a useful reference point (baseline) typical of a natural or little-altered river; and inability to fully address this issue of catchment connectivity. Where freshwater systems are fragmented, truly effective ecological restoration is difficult; at the same time, inappropriate catchment management can exacerbate issues like flooding whereby water is moved speedily through catchments with devastating effects on downstream communities (worldwide, the vast majority of catchments have their urban areas downstream rather than in the uplands).

Implicit in the many studies is the idea that human involvement in catchments equates to "damage." It is not so important when this damage began to occur (debate continues as to the timing of the onset of the Anthropocene, often seen as the point where rivers of prior reference status began to significantly degrade, with "degradation" broadly defined as ecological and physical simplification). According to Crutzen and Stoermer (2000), the start of the Industrial Revolution (ca. 1800) is the point when human activity accelerated so dramatically that humans became a dominant force on the planet and its water bodies. Kirch (2005), on the other hand, asserts that while human-induced changes to the global environment have accelerated with industrialization over the past three hundred years, such changes have a significantly longer cultural history, highlighting deforestation, spread of savannah, and rearrangement of landscapes for agriculture as examples. The conclusion of Waters et al. (2016) that the Anthropocene is functionally and stratigraphically distinct from the Holocene can only induce a stream of works each claiming to pinpoint the date of commencement of the new geological epoch. The time scale we adopt in this chapter is that of the "Great Acceleration" (Steffen, Crutzen, and McNeill 2007). Far more important overall is how we view and determine the *future* for rivers in the Anthropocene given that most rivers globally no longer operate according to their "Holocene norms" (Large and Gilvear 2015). Steffen, Crutzen, and McNeill (2007, 618) contend that that we are already at the beginning of the "third stage of the Anthropocene" (stages 1 and 2 being the Industrial Era and the Great Acceleration, respectively), where the "recognition that human activities are indeed affecting the structure and functioning of the Earth System as a whole (as opposed to local- and regional-scale environmental issues) is filtering through to decision-making at many levels." This growing awareness of human influence on the earth system has been aided by rapid advances in research and understanding. Pastore et al. (2010) emphasize the importance of understanding how humans have shaped the hydrology of the past in order to expand our understanding of the hydrology of today and of the future.

For rivers of the Anthropocene new organizational frameworks are needed for transdisciplinary investigation. These frameworks need to encompass the four areas referred to above, water engineering, land cover change, climate change, and human decision making, but also include the questions of motivation and impact. Researchers need to debate what constitutes "damage" and what it means to "restore" freshwater systems. What constitutes ecosystem "health" in the Anthropocene is also not at all clear, despite some notable international water legislation that has already been enacted (notably, the European Union Water Framework Directive at whose core is the definition of aquatic system health).

"Conservation-based management" and "design with nature" (Downs and Gregory 2004) have been identified as key approaches to the management of modern rivers. In both approaches there is an implicit reference to the current damaged state of river ecosystems and loss of reference status. This debate over "nature" is a vital component of both the scientific and popular agendas for sustainable development (Newson and Large 2006) but in extreme cases can become a barrier to efficient restoration projects. In some cases historical assumptions regarding "nature" can be confounding; Walter and Merritts (2008), for instance, highlight how a vision of an "ideal meandering form" exemplified by gravel-bed rivers has dominated restoration efforts in many riverscapes of the United States. In fact, the pre-European settlement of swampy landscapes and forest-dominated anabranching systems with cohesive sediments was the markedly different reality.

Effective management of Anthropocene riverscapes therefore requires more structured condition assessments. Where and what are the major riverine habitat areas under threat? Which are of greatest priority for river conservation, and why? What are their optimal sizes and spatial arrangements? What will be the effects of widely predicted global climate change? Globally, there is still an urgent need to effectively map refugia in order to boost chances of restoring key communities within catchments. Carpenter et al. (1992) have highlighted the potential impacts of global climate change on freshwater systems, and the United Nation's Millennium Ecosystem Assessment (MEA 2005) has clearly shown that in terms of drivers of change freshwaters in particular have experienced very high rates of habitat change and pollution and that these impacts are increasing at a rapid and worrying rate. While we cannot ignore these warning signs, geographically the type and scale of impacts also differ markedly. In the world's drier lands the main perceived fluvial damage is that caused by dams to flow regimes (Graf 1999; Newson, Pitlick, and Sear 2002). Elsewhere, the "damage" inflicted by flood defense works during the past century currently dominates the agenda of river restoration in northern Europe and North America. In all cases, management for the future of our Anthropocene rivers is complicated by the specter of climate change, with the current forecast being increased incidence of extreme drought and flood events in a warmer world (Kendon et al. 2014).

This takes us back to the issue of what constitutes a "natural" riverscape and to what extent this should actually constitute reference conditions in catchments that we cannot hope to return to their pre–Industrial Revolution status (fig. 2.1). Acreman et al. (2014) conclude that in heavily modified river systems lower expectations of a return to "naturalness" lead to flow regimes designed both to maximize natural capital and to incorporate broader socioeconomic benefits. Defining how far such rivers have shifted from their historical (i.e., dynamic) equilibrium requires extensive monitoring, which entails significant economic costs. Associated issues include (a) prevalence of suboptimal monitoring strategies, (b) an assumption of "active" engineering-based restoration (again costly in economic terms) rather than "assisted natural recovery" (Newson and Large 2006), and (c) a lack of evidence linking restoration/rehabilitation with tangible ecological and economic benefits. To assess rivers effectively so that our "Anthropocene management interventions" are deemed similarly effective, methods need to be developed that integrate river system hydrology/hydraulics, geomorphology, and ecology (and the complex interplay between these three different scientific disciplines). This leads to a challenge for scientists, policy makers, and managers of rivers as to how we can effectively merge quantitative models of earth systems and human systems with the more qualitative approaches prominent in the environmental humanities to establish effective baseline assessments. As Carpenter et al. (2009, 1305) conclude, "New research is needed that considers the full ensemble of processes and feedbacks, for a range of biophysical and social systems, to better

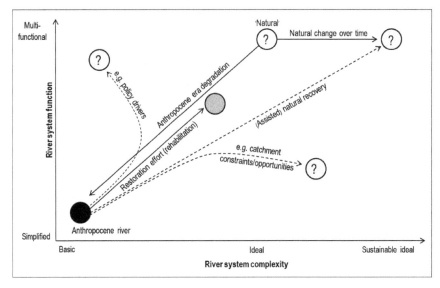

FIGURE 2.1. The response of river systems to anthropogenic drivers, illustrating shift from historical equilibrium conditions (degradation) and potentially different endpoints of restoration dependent of based on opportunities for, and constraints against improvement and wider policy drivers. The complicating factor of inherent/natural system change over time (also known as "shifting baseline syndrome") is also depicted; this will affect the vision for improvement. Modified from Bradshaw 1988.

understand and manage the dynamics of the relationship between humans and the ecosystems on which they rely."

According to Olsen (2002), if we know the baseline for a degraded river, we can work to restore it. But if the baseline shifts before it can be properly quantified, there is a danger we can end up accepting a degraded state as normal—or even as an improvement. The term "shifting baseline" was coined by Pauly (1995), who noted that each generation subconsciously views as "natural" the way in which their surroundings appeared in their youth. Although Pauly described shifting baselines in relation to fisheries science, the phenomenon is general and applies to all sectors of society. As one generation replaces another, people's perspectives change such that they fail to appreciate the extent and implications of past and current environmental modifications. Olsen (2002) provides an illustration of shifting environmental baselines in the Pacific Northwest's Columbia River, where the number of salmon in the river at the start of the twenty-first century and after an intensive effort at restoration was two times the population of the 1930s. In itself, that number is encouraging—but only if the 1930s numbers comprise the accepted reference point or baseline. In reality, salmon numbers in the Columbia River in the 1930s were only 10 percent of what they were in the 1800s, so, as Olsen (2002) points out, the 1930s numbers for the Colorado reflected a baseline that had already significantly shifted over the historic period. Papworth et al. (2009) present evidence for two distinct forms of shifting baseline syndrome: "personal amnesia," where knowledge extinction occurs as individuals forget their own experience, and "generational amnesia," where loss of knowledge occurs because younger generations are simply not aware of past conditions or baselines. This is reflected in figure 2.1, in which change over time (top) is associated with a loss of knowledge as to what type of system actually should represent the reference point in terms of what restoration outcome is deemed desirable or appropriate. Waldman (2010) recognizes this in stating that to put an end to the kind of persistent ecosystem degradation such as rivers and their watersheds have experienced, we will need to "rewind" important historical connections and interdependencies. Although it is important to look back for context, it is now more important to look forward to what society wants for our rivers in the future. Indeed, while it is important that we reestablish many of the connections and interdependencies of the past, we must also recognize that the watershed-scale fluvial processes that control the nature of our river environments can never again match those of the more undisturbed past. Therefore, there is a pressing need to understand "modern" aquatic ecosystem functioning and the constraints that our watershed usage imposes on the ways we manage our rivers in order to deliver the vital services to society that they wish for in the future. Waldman (2010) concludes that no less important in achieving this will be the tools, funding and legislation, and education to build social awareness and, crucially, the will on the part of politicians, policy makers, and the public to enact meaningful change.

TOWARD AN ECOSYSTEM SERVICE-BASED APPROACH

In 2000, then General Secretary of the United Nations Kofi Annan made a call for the first comprehensive assessment of the state of the global environment. The outcome was the Millennium Ecosystem Assessment (MEA 2005). Unsurprisingly, one of the key conclusions of the MEA was that over the preceding fifty years, humans, in the course of achieving substantial net gains in economic development and overall well-being, have degraded river ecosystems more rapidly and extensively than at any other time in history. This leads to two interconnected issues: while it is highly probable based on past evidence that ecosystem degradation will continue to worsen as we move deeper into this century, the challenge of reversing this degradation while meeting increased demands for "ecosystem services" (due primarily to population rise) will require major changes in institutions, policies, and practices. The 2005 MEA report uses a utilitarian definition of ecosystem services as the benefits people obtain from ecosystems (divided into "supporting," "provisioning," "regulating," and "cultural" services) and emphasizes the links between human well-being and these ecosystem services as being those of security, basic material for a good life, health, and good social relations. However, it should be recognized that ecosystem services, at least as defined here as qualities of ecosystems that benefit people, is not the same thing as an "ecosystem approach" to managing rivers. That distinction becomes important in discussing human modification of rivers and what it might mean to restore such rivers.

Worldwide, politicians, legislators, and policy makers are starting to recognize that aquatic systems comprise precious resources, not only providing the essentials of life—air, water, food, and fuel—but also underpinning national health, well-being, and prosperity and providing the potential for significantly improving quality of life. At the same time, it is increasingly understood that critical thresholds, or "tipping points," exist (Rockström et al. 2009a, 2009b; Biermann et al. 2012), beyond which sharp reductions in ecosystem service provision may result. In the United Kingdom, securing and maintaining a healthy natural environment and avoiding such thresholds is one of the government's two high-level goals, the other being tackling climate change. New ways of thinking and working have to be adopted for watersheds whereby the focus of policy making and delivery needs to be shifted away from isolated natural environment policies for air, water, soil, and biodiversity toward more holistic or integrated approaches based on whole ecosystems. Intrinsic to this shift are innovative yet widely accessible ways of assessing river system status and making this information widely available to a range of managers and interest groups. Such assessments of ecosystem services delivered by in riverscapes are starting to grow in number. The 2005 MEA provided the impetus in the United Kingdom for the 2011 UK National Ecosystem Assessment (UKNEA), the first major analysis of the nation's natural environment in terms of the benefits it provides to society and continuing economic prosperity. The UKNEA represented a wide-ranging, multi-stakeholder, cross-disciplinary (as opposed to transdisciplinary) undertaking. It was also aligned with other international initiatives, including The Economics of Ecosystems and Biodiversity (TEEB) study, a major international initiative whose findings were initially published in 2010, and the ongoing UNEP Intergovernmental Science-Policy Platform on Biodiversity and Ecosystem Services (IPBES). The UKNEA aimed to provide a comprehensive picture of past, present, and possible future trends in ecosystem services and their values (see, e.g., Brown et al. 2011), with one of the key objectives being to identify and understand what has driven the changes observed in freshwater systems and associated implications for ecosystem service delivery since 1950, the period coinciding with the Great Acceleration.

Despite the increasing number of assessment methodologies, there are a variety of reasons for the relative lack of impact of the MEA and associated national-scale initiatives. Prominent among these are (a) persistent gaps in the ecosystem services knowledge base, (b) a lack of operational tools and methodologies, and (c) limited awareness and understanding among decision makers of the concept of ecosystem services (fig. 2.2). While conceptual models of links between catchment landscape management, ecosystem services, and resultant human well-being exist, the scientific assumption of a direct link between geomorphic features and processes, ecological functions, and, for example, biodiversity remains largely

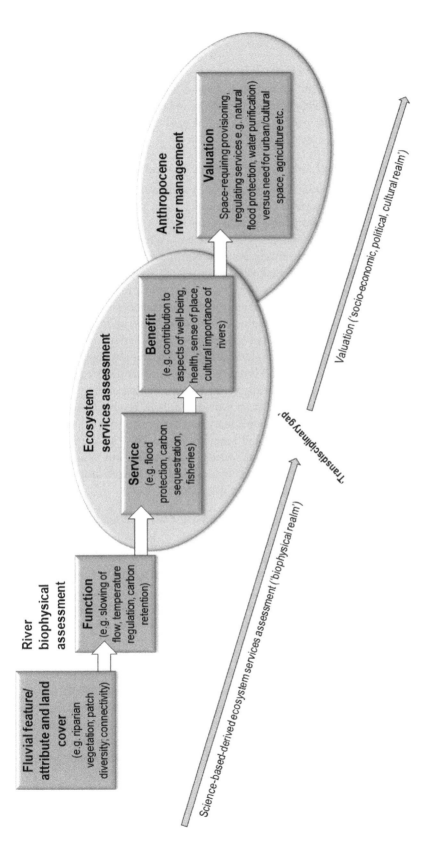

FIGURE 2.2. The cascade model of Haines-Young and Potschin (2010), emphasizing the transdisciplinary "gap" between science-based ecosystem assessment methodologies and the valuation of these ecosystems by society. More effective transdisciplinary frameworks are needed to educate society about the benefit of ecosystem services for underpinning wider social livelihoods.

unproven in either a systematic or a statistical sense. As mentioned above, it is imperative to merge our knowledge of earth systems and human systems with the qualitative approaches prominent in the environmental humanities to effectively value the benefits healthy ecosystems provide to society.

Rapid and Novel Assessment of Riverine Ecosystem Services

The importance of effective tools has already been emphasized above. In seeking to improve Anthropocene rivers, managers need specific tools to improve the information base on hydromorphological character and condition across entire sector lengths (i.e., upstream, mid-reach, downstream). While this is important to reduce issues introduced by shifting baseline syndrome, tipping points may vary from place to place within watersheds as some sectors may naturally be more robust than others. In addition, some sectors are more prone to anthropogenic alteration than others; for example, most large urban settlements are constructed in downstream reaches. While degradation of ecological integrity is typified by loss of landscape diversity, impairment of ecosystem function, and structural simplification, the relative importance of physical habitat degradation compared to other pressures (e.g., diffuse pollution) is not fully clear. Large and Gilvear (2015) emphasize therefore that any methodology aimed at quantifying or even simply defining the ecosystem services that rivers provide needs to be able to assess a "triple bottom line" of heterogeneity, connectivity, and dynamism both in a meaningful way and at appropriate spatial and temporal scales. The rapid uptake of remote sensing we have seen over the past decade for mapping and monitoring river status and health at multiple, often hierarchical, scales in the catchment context has potential value in assisting meaningful assessment.

Assessing Ecosystem Service Provision Using Virtual Globe Technologies

For a range of earth systems including freshwaters, Brown et al. (2016) conclude that we need to improve our criteria for diagnosing human impacts on the connectivity, integrity, and resilience of critical zone processes. Panoptic mapping tools including Google Earth and other virtual globes (e.g., Microsoft's Virtual Earth, NASA's World Wind) offer much potential for such assessments of rivers of the Anthropocene. A key advantage is that these mapping tools are free and easily accessible and offer global coverage of both heavily modified and less disturbed catchments. Potential users of these tools simply need the skill sets to identify relevant riverscape-scale features and the ability to extract riverscape features/attributes from remotely sensed data at appropriate scales. The outputs for managers and planners are science-based protocols for assigning riverscape features, or "attributes" to individual river ecosystem services within a robust, widely accessible metric-based system. Visualization tools like Google Earth can therefore help bridge the gap between researchers and those who need most to be reached with the results of research—policy makers and the population that lives in affected catchments.

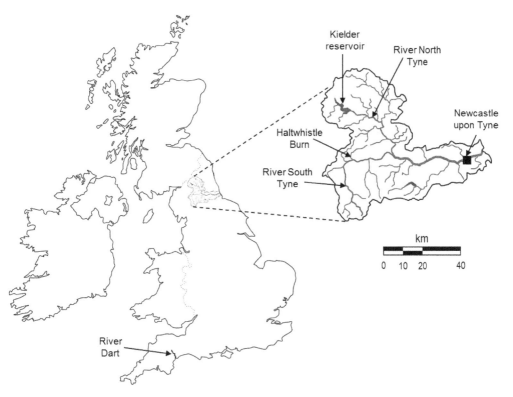

FIGURE 2.3. The River Tyne and the River Dart, U.K., showing locations referred to in the text.

The hydrological, geomorphological, and ecological linkages of water and sediment with biota within river systems drive the relationship between river processes, habitat provision, and ecosystem service delivery (Thorp, Thoms, and Delong 2006, 2008; Large and Gilvear 2015). Attributes of rivers that enhance heterogeneity, connectivity, and fluvial dynamics within river corridors enhance ecosystem service provisioning while at the same time being identifiable via remote sensing techniques. Efforts are under way (e.g., Large and Gilvear 2015) to develop Google Earth–based protocols for assessing the role of physical and biotic attributes initially on eight widely recognized ecosystem services. In the tool, Provisioning ecosystem services were defined as those of fisheries, agricultural crops, timber, and water supply; Regulating services were flood mitigation, carbon sequestration, and water quality control. In the first iteration of the tool, Supporting services were limited to the umbrella term "biodiversity." Cultural services were not specifically included, reflecting the difficulty of developing transdisciplinary assessments of rivers.

Here we apply the ecosystem services assessment tool to the River Tyne, United Kingdom. The Tyne (fig. 2.3) has two main tributaries, the North and the South

Tyne, and in total covers 2,933 square kilometers. The river has a mean annual discharge of 34m^3 s^{-1} (Jones, Lister, and Kostopoulou 2004) and is flanked by the North Pennine mountains to the west, the Cheviot Hills to the north, and the North Sea to the east (Environment Agency 2000). The majority of the population live in the Lower Tyne Valley, with the highest concentration in the urbanized east and coastal strip (Large and Gilvear 2015). While two-thirds of the catchment area is agricultural, and these activities have led to a mix of upland moor, forest, and arable land and pasture (Large and Gilvear, 2015), the Industrial Revolution—and particularly the coal-mining, ship-building, and heavy engineering industries—is central to the socioeconomic and cultural histories of the Tyne.

Figure 2.4 shows the resultant output of the ecosystem services assessment tool as a variety of indices, including a "feature/attribute" score and a "total ecosystem services" score. Two further indices can be calculated at the whole-river scale: a total individual ecosystem system score for each of the eight services identified and, summing these, a total ecosystem services score for the main river channel as a whole. Figure 2.4b compares the North Tyne, impounded in its upper reaches by Kielder Reservoir, to the less regulated River South Tyne (fig. 2.4a). Downstream from this confluence the traditional name "River Tyne" is used. The ecosystem service scores display a distinct "sawtooth" sequence, with "troughs" reflecting declines in ecosystem service delivery of the river at that specific point in the catchment. In the case of the Tyne, this ecosystem service decline is directly associated with human modification in the form of small urban centers, fluvial engineering in close proximity to the channel for transport infrastructure, impoundment in the form of Kielder Reservoir on the River North Tyne, and, importantly, the spatial footprint of the city of Newcastle upon Tyne toward the downstream end of the main river channel. Figure 2.4c describes an alternative scenario, that of increase in ecosystem service delivery for the River Dart in Devon, U.K., where sedimentation following decommissioning of an in-channel weir resulted in the (unintended) consequence of ecological improvement via alluvial woodland formation (F in fig. 2.4c).

The tool has already been taken up in the United Kingdom by several non-governmental bodies, including River Trusts, so it would appear a Google Earth assessment based on identifiable fluvial attributes is relevant to a wide sector of users, planners, scientists, and the general public. It is hoped that with development of the approach it can (a) highlight unintended consequences of actions in river systems, (b) evidence shifting baselines affecting conservation management and restoration, and (c) effectively demonstrate opportunities for win-win synergies between environmental management disciplines in specific parts of catchments where optimization of ecosystem service delivery is a desired objective (Everard and McInnes 2013). In the Tyne and other U.K. catchments, reorienting the EU Water Framework Directive goals of "good ecological status" toward maximized ecosystem service provision can potentially deliver greater societal benefit within multiple-use river landscapes (Stanford and Poole 1996; Everard 2011).

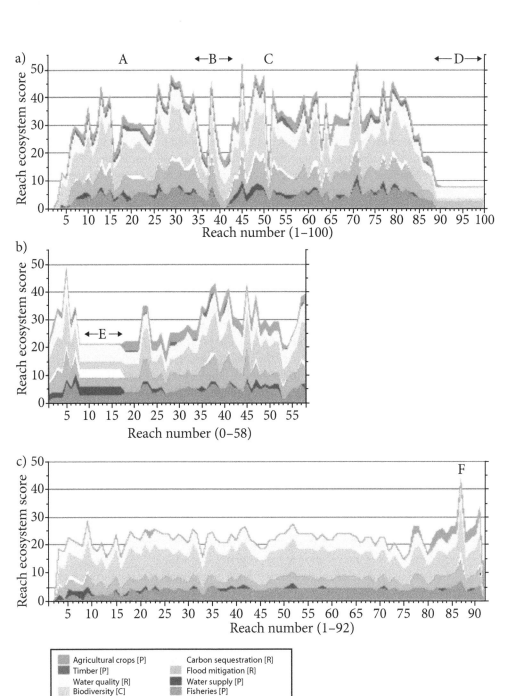

FIGURE 2.4. Downstream patterns in ecosystem service scores and total ecosystem service scores based on Google Earth assessment of ecosystem services from fluvial features (Large and Gilvear 2015). (a) River South Tyne and River Tyne, (b) River North Tyne to its confluence with the South Tyne, (c) River Dart, Devon, U.K. A–D: decline in reach ecosystem service score due to urban settlements of Alston (A), Haltwhistle (B), Haydon Bridge (C) and Newcastle upon Tyne (D); E: Kielder Reservoir. F: increase in ecosystem service provision due to localized sedimentation following decommission of a weir on the River Dart.

TOWARD WATERSHED-SCALE TRANSDISCIPLINARITY

What constitutes a healthy Anthropocene river? As numerous researchers (e.g., Vorosmarty et al. 2010; Carpenter et al. 2011; Sponseller et al. 2013) have attested, the multiple roles that water plays in both minimally and intensively manipulated ecosystems raise numerous challenges for efforts to reverse degradation. Perhaps most problematic is the lack of truly integrative approaches linking "supporting," "provisioning," and "regulating" ecosystem services with "cultural" ecosystem services. Despite this, the ecosystem service concept resonates with river scientists as it emphasises the need for healthy system structure and function, while at the same time concentrating attention on what makes rivers so valuable to human society. Rivers in the Anthropocene offer a range of services beyond those underpinned by ecological diversity; even simplified urban rivers still provide ecosystem services and can still function as havens of tranquillity and meaning. Most reference scenarios for restoration have some pristine view on what a channel should look like at their centre, despite the fact that wilderness channels are not the most beneficial for humans. In Scotland, a scheme run by WWF in the mid-1990s called "Wild Rivers" faltered as a result of the public's negative perception of the term "wild." Elsewhere, many rivers under intense human pressure have huge value in terms of their religious significance and often sacred nature, while recreation is also a major user of the world's freshwater systems.

Finding ways of properly integrating these socioeconomic and sociocultural aspects with more traditional life science and geomorphological approaches to ecosystem service-based management is a fundamental need as we move further into the Anthropocene. Numerous issues remain to be addressed (table 2.1). For effective and cost-beneficial restoration, managers of rivers like the Tyne need to know with what aspects of physical habitat and at which locations in catchments intervention will lead to the greatest improvements in ecological condition and protection/enhancement of ecosystem service delivery. Managers also need to know what *kinds* of intervention are appropriate, and *where*.

In terms of the first issue listed in table 2.1, it is undeniable that while patchiness is awkward to manage, the patch mosaics arising from heterogeneity, connectivity, and dynamism are essential to riverscape-scale (Thorp, Thoms, and Delong 2006, 2008) relationships between fluvial features, land cover types, natural ecosystem functions, and river ecosystem service delivery (fig. 2.5). Given that the riverscape is where people live, Anthropocene river management requires improved understanding of these relationships between people and the physical system they inhabit and its natural and cultural ecology. Carpenter et al. (2009) have described this as a need for improved understanding, which can only come from enhanced knowledge transfer between environmental scientists, geographers, social scientists, industry practitioners, and the general public (see also Newson and Large 2006). Key to managing degraded Anthropocene rivers worldwide is greater appreciation in a range of communities for the value of heterogeneity, connectivity,

TABLE 2.1. Persisting Management Issues Associated with Anthropocene Rivers

Issue	Details
Management	Rivers and streams are individually unique, patchy, discontinuous, and strongly hierarchical systems [*i.e., they are awkward to manage*].
Data gaps	Existing applied approaches for capturing geomorphological data are highly dependent on intensive fieldwork and monitoring; this is unlikely to be resourced at sufficiently extensive scales to meet management needs. While a large number of hydrometric monitoring stations have gone out of service since the mid-1980s, data-gathering capacity from remote-sensing has increased almost exponentially.
Habitats and flows	Fully integrated ecosystem service-based approaches/assessments are not yet operational, and there is a pressing need to collate data to describe the habitats on which biotic function depends over the full range of flows.
Classification	Robust, process-based geomorphological typologies incorporating full dynamic assessment over the whole flow regime are needed to better define river variability [*currently impracticable due to lack of communication between individual disciplines*].
Modification	Reliable spatial data (and maps) of channel modification are lacking [*need for enhanced communication of scientific needs and data requirements to politicians, managers, and the general public*].
Hydrological variability	Little attention is paid to natural hydrological variability within riverine systems, despite the fact that this factor has a defining role in the hydromorphology of the system [*hard to manage running water systems without this scientific insight*].
Shifting baselines	Problematic for those using human perceptions of change to inform river policy making or management.

and dynamism in the landscape. Natural features of rivers interacting with natural flow dynamics positively enhance heterogeneity, connectivity, and fluvial dynamics within river corridors and in turn enhance ecosystem service provisioning (Large and Gilvear 2015). On the other hand, human modifications that simplify or degrade these attributes have tended to simplify river ecosystems and degrade ecosystem service delivery, or what nature does for us, with the main exception being increased supply of products from manipulated river systems (timber, fisheries, water supply from impoundments, etc.).

Riparian Communities and the Growth of "Crowd-Sourcing"

In 2011, the U.K. Department for Environment, Food, and Rural Affairs (DEFRA) announced a reemphasis on a catchment or watershed-based approach to restoration. The stated vision was to "provide a clear understanding of the issues in the catchment, involve local communities in decision-making by sharing evidence, listening to their ideas, working out priorities for action and seeking to deliver integrated actions that address local issues in a cost effective way and protect local resources" (Defra 2011). This mandate clearly also applies to the River Tyne but was perhaps more succinctly described by the Tyne Rivers Trust (2012) as "action to improve our rivers, and action to raise awareness and educate people about

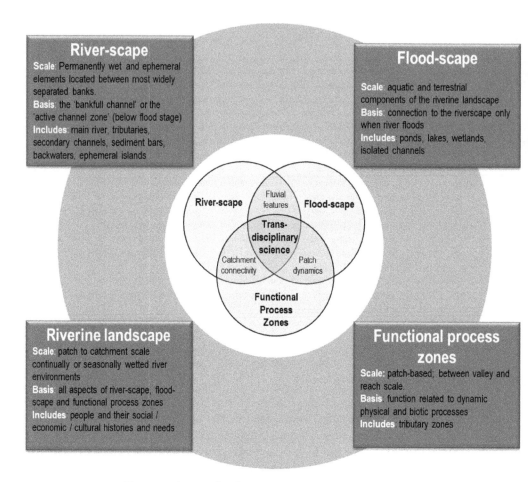

River-scape
Scale: Permanently wet and ephemeral elements located between most widely separated banks.
Basis: the 'bankfull channel' or the 'active channel zone' (below flood stage)
Includes: main river, tributaries, secondary channels, sediment bars, backwaters, ephemeral islands

Flood-scape
Scale: aquatic and terrestrial components of the riverine landscape
Basis: connection to the riverscape only when river floods
Includes: ponds, lakes, wetlands, isolated channels

Riverine landscape
Scale: patch to catchment scale continually or seasonally wetted river environments
Basis: all aspects of river-scape, flood-scape and functional process zones
Includes: people and their social / economic / cultural histories and needs

Functional process zones
Scale: patch-based; between valley and reach scale.
Basis: function related to dynamic physical and biotic processes
Includes: tributary zones

River-scape — Fluvial features — Flood-scape
Trans-disciplinary science
Catchment connectivity — Patch dynamics
Functional Process Zones

FIGURE 2.5. The terminology used in the riverine ecosystem synthesis of Thorp et al. (2006, 2008) adapted to show the potential central role for transdisciplinary river science.

the importance of rivers." The Rivers Trusts are relatively small environmental charities entrusted by the U.K. government to produce plans for whole catchments, which in the case of the Tyne entails an area of almost 3,000 km². In the United Kingdom organizations like the Rivers Trusts are now deemed essential to achieving action on the ground, via what the Tyne Rivers Trust (2012) refers to as "perpetual partnerships," helping to offset the personal and generational amnesia associated with the evolution of our river histories. In their *River Tyne Catchment Plan* published in December 2012, the Tyne Rivers Trust produced a publicly informed "wish list" of proposed projects that aims to deliver better rivers within the wider Tyne catchment and to increase community involvement in local decision making. The intention is to engage and educate those who are not aware of the importance of rivers, create robust and resilient watersheds to cope with weather extremes and climate change, and make the best use of all available resources, research, and evidence to support work across the catchment and deliver the targets set out in legislation like the EU Water Framework Directive and the EU Habitats Directive (Tyne Rivers Trust 2012). Of importance here is that when the United Kingdom leaves the European Union similar legislation will be enacted to succeed the Directives.

As an example of this new consultative approach in action, Newcastle University has worked in close partnership with the Tyne Rivers Trust on a project focused on Haltwhistle Burn, a small (42 km^2) rural subcatchment of the River Tyne (Starkey and Parkin 2015; Starkey et al. 2017). Funded by the U.K. government's Catchment Restoration Funds Project and the Natural Environment Research Council, this "total catchment approach" seeks to improve fish populations, water quality, and hydromorphology and reduce flood risk. The major objective in establishing future priorities for the catchment is to engage with the local community by using established natural runoff management, with the ultimate aim of producing a catchment management plan for Haltwhistle Burn. The involvement of local communities in knowledge production will avoid the pitfalls of shifting baseline syndrome. It is intended that the experiences gained during the Haltwhistle Burn project will be transferred to other Anthropocene watersheds where Rivers Trusts are responsible for on-the-ground management via partnership with local communities. The ambition is to maximize the size of catchments addressed; science-based approaches have traditionally only been aimed at relatively small experimental catchments.

Figure 2.6 illustrates observations submitted by local residents in early 2014, as a result of efforts to encourage a "community monitoring" approach. If we are to advocate "citizen science" as a component of a transdisciplinary approach to Anthropocene catchment management, there are a number of considerations to address. What are the key participant needs and motivations for engagement and recruitment? What training and data collection resources are required to ensure good quality and consistent observations? How should the data gathered be managed, analyzed, visualized, disseminated, and shared? What are the key ethical and social, economic, and practical considerations? For maximizing the sustainability and legacy of a citizen science or "volunteered geographical information" project such as that at Haltwhistle Burn, a key objective is finding a way to encourage volunteers to carry the process into the future. Findings from the Haltwhistle Burn indicate it is best to engage on a local level and to ensure citizen science is related to a relevant topic or issue that directly affects riparian communities (e.g., flooding), with findings constantly fed back to the community using effective visualization tools such as the annotated hydrograph in figure 2.6. While citizen science = knowledge coproduction = new power, citizen science is not just about knowledge coproduction; the project can, and should, aim to have a variety of social benefits. For example, one citizen scientist from the Haltwhistle community stated, "I'm starting to understand the wider picture," and another said, "I'm really getting into

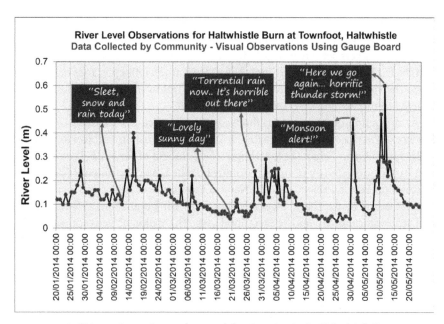

FIGURE 2.6. Citizen science via crowd-sourced data in action in the Haltwhistle Burn catchment. Local community river level and weather observations were collated via the social media platform of Twitter (@HaltwhistleBurn) and a purpose-built "community river, weather and flood" Android mobile phone app.

this science stuff." Using volunteers, information and data can be gathered over a wide area. While "any data is better than no data," it is vital to maximize the credibility of citizen science observations, and therefore protocols are needed to limit error and uncertainty and create metadata (i.e., information or data that explain the data). One of the biggest challenges associated with a citizen science approach is getting professional scientists to accept, appreciate, and actually use the data to support decision making and to underpin evidence-based policy (European Commission 2013). If this can be achieved, there are a wide range of potential applications for this type of data, including catchment modeling and flood warning schemes, as well as ongoing monitoring of natural flood management initiatives.

CONCLUSION: IMMEDIATE PRIORITIES FOR ANTHROPOCENE RIVER MANAGEMENT

Recognizing the advent of the Anthropocene raises challenges for how we perceive our rivers should behave. It also raises challenges in terms of how we can ensure their sustainable futures. In January 2014, a transdisciplinary workshop, "Rivers of the Anthropocene," held at IUPUI in Indianapolis, brought forward a number of important questions. How do scholars from different disciplines frame the problems of environmental change differently? In what ways does a transdisciplinary

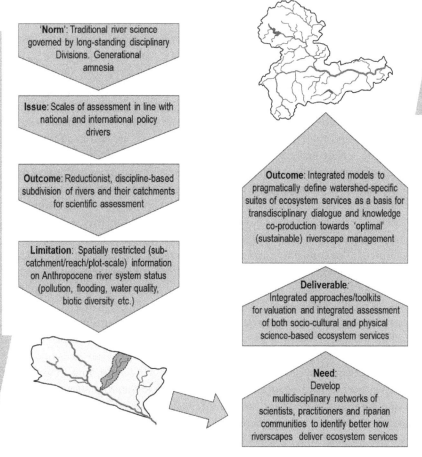

FIGURE 2.7. Issues and limitations of "traditional" river science with its often reductionist outcomes and the positive deliverables and outcomes offered by transdisciplinary approaches aimed at better integration of sociocultural and physical science–based ecosystem services.

perspective alter their approach? What problems does it create, and what are the most effective ways to solve them? How can we reframe ideas and approaches that are embedded in traditional disciplinary constructs? In seeking answers to these questions, we need models of human-environment interaction that account for both emergent environmental phenomena and the agency of human societies. The challenge is to make these meaningful in terms of multiple scales (time, population, geography), forms of flow (interaction, feedback), and properties of change (emergence, agency, rate, cause and effect). In coproducing scientific knowledge on these models it is vital to engage with as wide a range of user, practitioner, and academic communities as possible, in order to develop new, transdisciplinary approaches based on riverscape ecosystem services (fig. 2.7). Google Earth and other freely available virtual globes offer a great deal of potential, as do frameworks such as the ecosystem services "cascade model" of Haines-Young and Potschin (2010).

For better management of our Anthropocene river systems we need to advance appreciation of how habitat features in riverscapes underpin ecosystem service provision. This exercise should aim to reduce the need for "expert judgment" to determine what constitutes "optimal" ecosystem service delivery. Approaches using science-based tools run the risk of lower uptake in more populated watersheds, where system dynamism is seen as an inherent threat rather than a mechanism to ensure sustained ecosystem service delivery. There is a need to involve stakeholders, policy makers, and the general public in knowledge production to increase our understanding of how both more pristine and more intensively used riverscapes deliver ecosystem services in their own right; both simple and more complex habitat types can ultimately deliver similar levels of societal benefits.

Carpenter et al. (2009) point out that while sustainability science is motivated as much by fundamental questions about interactions of society with its surrounding environment as by compelling and urgent social needs, many aspects are currently based on assumptions rather than data. For example, one of the biggest issues in assessing the implications of shifting baselines syndrome is a lack of empirical evidence that it actually occurs (Papworth et al. 2009). Carpenter et al. (2009) advocate expanding basic research on social-ecological systems and building on disciplinary strengths while at the same time bridging disciplinary divides to create the new knowledges needed to build our Anthropocene watersheds into resilient social-ecological systems. Protocols linking relevant science with an informed public have been advocated for some time. For example, Stanford and Poole (1996) describe iterative protocols for involving scientific research and public opinion in adaptive ecosystem management, and Ostrom (2009) advanced a generalized framework for analyzing sustainability of socioeconomic systems. Thus far, however, there has been difficulty in assessing cultural ecosystem services (Schaich et al. 2010), and an immediate priority must be finding ways of effectively integrating cultural ecosystem services with supporting and provisioning ecosystem services in rapid assessment methodologies. Allied to this is the need for widely accessible decision-making tools and guidelines that implicitly recognize societal valuation of ecosystem services in terms of what nature still does for us in our Anthropocene river systems.

3

Political Ecology in the Anthropocene

A Case Study of Irrigation Management in the Blue Nile Basin

Sina Marx

The study of human-environment relations and the focus on resource management practices have a long tradition in anthropological research. Early accounts that explain societies by means of their natural surroundings were based on a conceptual dichotomy between nature and society. Focusing on differences between societies (a) and the environment (b) was framed as the explanatory variable. The explanation of their connection was mostly one of a simple cause-effect relationship (a◊b), a framework that dominated the field far into the nineteenth century (Dove and Carpenter 2008, 1).

In the twentieth century, environmental determinism started to be increasingly contested within anthropology, and a more complex understanding of the relationship between societies and the environment developed. The focus shifted "to the asking of the reverse question, not how does the environment affect society but how, over time, does human activity affect, and especially degrade, the environment" (Dove and Carpenter 2008, 2). Further, with increasing globalization, the focus on local-level analyses alone became insufficient, and it was acknowledged that cultural as well as ecological processes on the local level were part of a broader set of both political and economic factors (Peet and Watts 1993, 1996; Bryant and Bailey 1997).

The paradigm shift to "the reverse question," how humans affect the environment, has largely been a story of degradation. The narrative of the 1960s proposed that population growth coupled with mismanagement of natural resources by local communities was the primary cause of environmental degradation in the so-called Third World (Neumann 2005, 26 f.).[1] Closely linked with such neo-Malthusian thinking was the notion that the needs of a growing population could be accommodated through technical and managerial improvements. As a consequence, classical development approaches aimed at implementing an agenda of such adjustments in Third World countries to overcome their environmental problems with the help of (Western) specialists in civil engineering or agronomy. Countering technocentric assumptions about the dynamics of resource use and environmental degradation and criticizing the neglect of social, economic, and political structures gave direction to early political ecologists' writings. Focusing on the social, economic, and political circumstances under which environmental conditions evolved worked to denaturalize nature, showing those conditions to be the outcomes of negotiated power relations (Blaikie and Brookfield 1987).

DEGRADATION GOING GLOBAL—THE ANTHROPOCENE

Today's narrative on anthropogenic changes to the earth system, the Anthropocene, has taken the discourse of degradation to a new, global level, which provides a rich field for studies in political ecology. The term "Anthropocene" represents the notion that through human actions, we have entered "a new phase in the history of both humankind and of the earth, when natural forces and human forces [have become] intertwined, so that the fate of one determines the fate of the other" (Zalasiewicz et al. 2010, 2231). This statement illustrates the conceptual scientific challenge of the Anthropocene: it breaks down what Chakrabarty (2012, 10) calls the "wall of separation between natural and human histories that was erected in early modernity."

However, framing humankind as a global force and calling the changes resulting from collective human actions "anthropogenic" implies a false homogeneity. It conceals the fact that neither the drivers nor the effects of global changes are equally distributed among humankind. That is why Malm and Hornborg (2014) call global environmental changes "sociogenic," rather than anthropogenic, as a way "to indicate that the driving forces derive from a specific social structure, rather than a species-wide trait" (6). When considering climate change as one effect of an economy that is based on fossil fuels, "humankind" and "anthropogenic" are not categories that can account for the fact that those who are least responsible for climate change in terms of carbon emissions are likely to suffer most from its impacts (fig. 3.1).

Humankind is obviously not a single global agent, particularly when one takes into consideration the inequalities between regions as well as those that exist within a country. Hence, "species-thinking on climate change is conducive to mystification and political paralysis. It cannot serve as a basis for challenging the vested interests of business-as-usual" (Malm and Hornborg 2014, 6).

Against this background, I explore how global institutions and discourses that evolved around the issue of anthropogenic global environmental changes modify

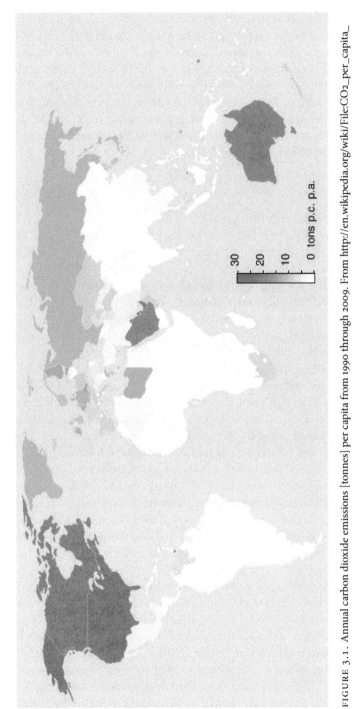

FIGURE 3.1. Annual carbon dioxide emissions [tonnes] per capita from 1990 through 2009. From http://en.wikipedia.org/wiki/File:CO2_per_capita_ per_country.png. Licensed under the Creative Commons Attribution-Share Alike 3.0 Unported license.

the agency of stakeholders at different levels in water management, using the example of the Koga project—the first large-scale irrigation scheme to become operational in the Ethiopian Blue Nile basin since the 1970s. The study looks at interstate relationships among riparians of the Nile basin under climate change—as the most prominent component of the Anthropocene—and the role of the Koga project in this context. The responses of ministerial actors to changing paradigms of resource management are described at the national level. Finally, the study explores the effects of such changing paradigms and actor constellations on local irrigation management.

METHODS

Fieldwork was conducted to examine the social impact of the irrigation scheme in two ways. It was done by looking, on the one hand, at institutional and organizational transformations and, on the other, at changes in the livelihoods of farming households affected by the project.

The qualitative methods applied during the research consisted mainly of structured and semistructured interviews, informal interviews, focus group discussions (FGDs), participant observation, and various rapid rural appraisal methods such as ranking exercises and transect walks. The interviews with farmers consisted first of a structured portion in the form of a household survey on the social and economic situation, including information about household members; cultivated crops; other means of income; distance to the nearest freshwater source; access to electricity, sanitation, health, and agricultural services; the farmers' organizations in place, and so on. The second part varied according to the answers given in the fisrt part of the interview and aimed at the perceptions of irrigation benefits and costs, technical problems, and social conflicts related to irrigation. Informal interviews and FGDs were conducted with staff from the engineering company responsible for the supervision and overall management of construction, as well as with those from the capacity building team that conducted trainings with farmers. A couple of these trainings were attended to conduct participant observation, summing up to a total of six full days of lectures and discussions and one day of training on the fields. Those farmers who received irrigation water for their fields had to become members of a water user association (WUA). I attended several meetings of the WUA's different bodies (five zonal meetings and one board meeting). Informal interviews were also held with staff from the government agencies involved, namely, the Ministry of Water Resources (MoWR) in Addis Ababa and Bahir Dar, the Bureau for Agriculture and Rural Development (BoARD) in Addis Ababa and Bahir Dar, and the local Cooperative Promotion Bureau (CPB) in Merawi, as well as with staff and customers of the Agricultural Service and Credit Service Cooperatives in Merawi.

In addition to the fieldwork, an extensive amount of the available project documents were reviewed and analyzed, including the Feasibility Study and its appendixes (AIL 1995a, 1995b), the Appraisal Report from the African Development Bank (2010), monthly and quarterly reports issued by the engineering companies from June 2003 on, and reports relating to Koga prepared by several consultant firms on behalf of the MoWR.

HISTORICAL CONTEXT OF IRRIGATION IN THE TRANSBOUNDARY NILE BASIN

Historically, irrigation has always been closely related to the formation of states and the exercise of power over its citizens (Wittfogel 1957). While large irrigation systems do not inevitably demand centralized authoritative management (Ostrom 1992, 1993; Shivakoti and Ostrom 2002), "regardless of the direction in which causality runs, harnessing water on a large scale has been associated with the formation of many early powerful states" (Barker and Molle 2004, 8). Irrigation has had a long history in Ethiopia, around two thousand years, albeit predominantly practiced on a small scale. With the coming to power of the Derg regime in 1974, irrigation—and in fact agriculture as a whole—declined quickly due to the ensuing socialist land reforms. Today only around 5 percent of Ethiopia's water resources are being utilized, so state intervention for irrigation on a large scale is rather at a beginning stage, with dam construction on the rise. Considering the social and environmental problems that large-scale projects in the water sector have caused in the past, "the new rush into large-scale irrigation is inviting a number of problems" (Moges et al. 2010, 83) that have already been recognized in the debates on dams of the 1990s and those on irrigation failure in Africa as a whole.

WATER STORAGE AND THE STRUGGLE OVER THE NILE

The seemingly paradoxical situation of about 110 billion m^3 of water flowing across the country's borders every year while a majority of the population lives in a state of constant undersupply of water is a result of both the high variability in rainfall and the lack of infrastructure. Because smallholders account for nearly 90 percent of the overall agricultural production in Ethiopia, and at the same time represent the group most vulnerable to uncertain climatic conditions, national food security is accordingly low.

However, while both the stakes and the potential for water storage are high, so too is the potential for disputes. Approximately 90 percent of the country's freshwater crosses international borders. Transboundary management of the resource is indispensable—with the Blue Nile (called the Abbay River in Ethiopia) being the most controversial. While the basin has an estimated irrigation potential of about 711,000 hectares (ha) (Arsano and Tamrat 2005), it is also the largest tributary to the Nile and is therefore subject to conflictive political and economic interests of the other riparian nations, especially those of Egypt and Sudan.

TABLE 3.1. Irrigation and Economic Indicators of Ethiopia and Egypt

Indicators	Ethiopia	Egypt
Irrigated land as % of total cultivated area[2]	2.5	100
Irrigated land in ha[2]	289,530	3,422,178
of which located in the Nile Basin[3]	76,000	3,080,000
Water withdrawal rate (m³/capita/year)[1]	48	1202
Employment in agricultural sector (%) in 2005[2]	81	31
GNI per capita in 2009 (US$)[4]	380	2,980

SOURCES: Compiled from [1]Gebeyehu 2004; [2]FAO 2005; [3]Kloos and Legesse 2010; [4]World Bank 2014.

Thomas Homer-Dixon (1994) has argued that in transboundary water management "conflict is most probable when a downstream riparian—a river-bordering state—is highly dependent on river water and is strong in comparison to upstream riparians." This is exactly the case in the Nile basin given the vast differences between the countries in use of water resources and economic indicators (table 3.1).

The area along the Nile in Egypt and Sudan is one of the largest contiguous regions of high irrigation density in the world, and Egypt—as the downstream riparian—is by far the most economically powerful. With around 63.8 cubic kilometers, the total water withdrawal in Egypt equals 3,794 percent of the internally available renewable water resources (FAO 2005, 63), and the vast majority of this water is taken from the Nile. Such inequalities have a long history and are entwined with the history of control over the Nile waters.

From the beginning of agriculture in the region of Egypt and Sudan, around five millennia ago, the Nile has been the basis of life for most of the area's inhabitants. About two thousand years later, artificial irrigation started, but it was not until the colonial interference of the British that Egypt began to systematically build dams and barrages. In 1929, Sudan and Egypt signed the first treaty exclusively dealing with the allocation of Nile water, allotting 48 billion cubic meters to Egypt and 4 billion to Sudan (Swain 2002, 296). After a phase of heightened political tensions due to the unequal distribution, the negotiations resumed, and in 1959 a new agreement was reached that assigned the entire average annual flow of the Nile to be shared between Egypt and Sudan—neglecting the rights of the remaining eight riparians. Ethiopia was allocated none of the Nile's resources, although it contributes 80 percent of the total annual flow.

Egypt has long since been unwilling to change the state of affairs by any form of cooperative management. While following its unilateral goals and projects on the Nile, Egypt has historically tried to prevent any upstream development to preserve its own control. However, since the beginning of the 1990s, Ethiopia has started to become a threat to Egypt's water supply as the country has begun its own irrigation projects on the Blue Nile. Despite protests by Egypt and Sudan, Ethiopia has insisted on its sovereign right to make use of the resources within its borders. The quarrel reached its peak when Egypt managed to prevent the African Development Bank (the very bank that financed the Koga dam project under study here) from financing Ethiopia's planned water projects (Swain 2002, 298). After decades of political tensions over the use of the Nile water, the establishment of the Nile Basin Initiative in 1999 represents the most promising attempt at basinwide cooperation to date. Even though "there is not yet a new water management regime in the basin, . . . Ethiopia continues to develop its bargaining power vis-à-vis its downstream neighbours and within the Nile Basin Initiative" (Cascão 2008, 27).

THE KOGA IRRIGATION SCHEME IN THE CONTEXT OF HYDROPOLITICS AND CLIMATE CHANGE

Against this background, the Koga project is an important experiment within the national Integrated Water Resources Management (IWRM) portfolio. Thus, it has gained wide attention not only in Ethiopia but on the international level as well, even making it into the *Wall Street Journal* in 2003: "The Koga River project is being cast as a 'confidence builder' to show that upstream uses don't necessarily hurt downstream populations. Ethiopian engineers calculate the Koga irrigation would use less than one-tenth of 1% of the Nile flow reaching the Ethiopia-Sudan border" (cited in Haileselassie et al. 2009, 132). Also keeping the challenges of transboundary water management in mind, its success will be crucial for further development of the sector as "achieving implementation targets will be viewed by the international community as an indication of Ethiopia's capacity to handle similar capital-intensive schemes in the future. . . . [It is] regarded by lending organizations as the nation's litmus test to successfully bargain and attract major loans for future investment in the Nile Basin" (Gebre, Getachew, andMcCartney 2007, 25). The political relevance of the project on an i interstate level is clear: it will affect Ethiopia's bargaining position within the hydropolitics of the Nile, while the legitimacy of infrastructure development increases with the spread of climate change policies.

For the research area in the highlands of Amhara Regional State, climate change scenarios suggest a probability of increased rainfall that could benefit crop yields and thus food security (see, e.g., Bates et al. 2008; Kim 2008). Increased drought is not one of the probable effects of climate change in the Blue Nile basin. Irrigation is still incorporated in the Ethiopian National Adaptation Programme of Action (NAPA) as one of the most important adjustments in the agricultural sector to ensure food security (MoWR, NMA 2007; Ludi 2009, 6).[2] The main reason for this, according to the document itself, is that "current climate variability is already imposing a significant challenge to Ethiopia by affecting food security, water and energy supply, poverty reduction and sustainable development efforts, as well as by causing natural resource degradation and natural disasters. For example the impacts of past droughts such as those of 1972/73, 1984 and 2002/03 are still fresh in the memories of many Ethiopians" (MoWR, NMA 2007). The AfDB Appraisal Report on the Koga project states the same rationale behind the Koga project: "The GOE [Government of Ethiopia] decision to accord the project a priority stems from frequent drought and food shortages" (AfDB 2001).

The problem of food insecurity can now be tackled with financial support from the international community that might not have been available without the climate change discourse. These developments allow Ethiopia to place water storage, as a national adaptation strategy, on the agenda despite the resistance from its downstream neighbors, a policy that otherwise might have been too politically sensitive to address with regards to interstate hydropolitics. However, as Lautze and Maxwell (2007, 239) point out, vulnerability to drought in Ethiopia "is known to arise from political marginalization rather than either technical deficiencies or the vagaries of the weather." The following two sections analyze in depth how far the Koga irrigation scheme, as a supposed technical solution, has succeeded in reducing this vulnerability.

MINISTERIAL POWER RELATIONS AND THE "NEED" FOR WATER USER ASSOCIATIONS IN IWRM

The ministries involved in policy making regarding water storage for food security are mainly the Ministry of Water Resources (MoWR) and the Ministry of Agriculture and Rural Development (MoARD), since irrigated agriculture is located at the interface of their responsibilities. The Ethiopian National Water Resources Strategy states that it is "promoting the principles of integrated water resources management" (MoWR 2001, 2), also attempting to mitigate the expected effects of climate change. Irrigation management transfer is increasingly promoted as a tool to manage demand in IWRM to both reduce costs and increase participation. In line with more general structural adjustment programmes starting in the 1980s, irrigation management transfer as one form of privatization has been supported by many of the major international development banks (FAO 2001; cf. EDI 1996). However, the form that management transfer can take varies greatly from scheme to scheme.

While by its design the Koga project was envisioned as the first large-scale irrigation scheme to be managed by the farmers themselves, inconsistencies arose during the implementation phase concerning what parts of the scheme the farmers were actually going to manage and what parts should remain under state responsibility. Interpretation of the envisioned "self-management by the beneficiaries" has been inconsistent and undergone a number of changes that can be tracked via the rich project documentation. While in 2001 the division of management duties was outlined in spatial terms (infrastructure down to secondary canals managed by experts; infrastructure up to secondary level managed by farmers), the entire responsibility and duty of management and operation were only ascribed to the beneficiaries in 2004 by the Cooperative Promotion Bureau (MMD 2005) in the course of establishing an irrigation cooperative (IC). Then the organizational framework was changed back to the initial plan of a jointly managed scheme intending to rely on the professional Project Management Unit to take care of the primary and secondary structures and support the nonprofessional IC in fulfilling the remaining duties. The legal status of the IC, however, remains unspecific. The title was usually applied by farmers and officials in an undifferentiated way from Water Users' Association, which is the form of farmers' organization put forward by the Ministry of Water Resources. Similar to other case studies, "no institution like the WUA formally exists. However, farmers mention them[;] . . . they claim to be a member of it" (Leidreiter 2010, referring to West Belisa). The "nonexistence" of WUAs is due to the fact that in Ethiopia the term usually refers to groups of farmers who organize irrigation themselves without official registration, while cooperatives are legally recognized by the Cooperative Societies Proclamation No. 147/1998. These nonprofit WUAs focus solely on water distribution, management, and operation of the infrastructure, but are "sometimes threatened by parallel established government-supported cooperatives which have broader operational scopes and have stronger links with government institutions" (Haileselassie et al. 2008, 35). However in the Koga case, both the existing IC and a potential WUA would be government-installed rather than driven by farmers.

Donors have contributed to this conceptual and legal confusion as well since they have imposed the internationally established concept of the WUA. As the World Bank stated with regard to the Ethiopian Nile Irrigation and Drainage Project, "Water users in Ethiopia have so far been mostly organised into legally recognized Water Users Cooperatives. . . . The project will sensitize communities on WUAs and encourage the formation of these in view of the comparative advantages as demonstrated in other countries" (World Bank 2007, 61). In 2009, the World Bank published a draft for the proclamation of WUAs, as well as for the establishment of by-laws and contract agreements, "to assist the Government of Ethiopia in the definition and adoption of the legal framework for the establishment of Agricultural Water Users Associations for the sustainable development and management of irrigation and drainage infrastructure" (BRLI 2009, 1).

This situation leads to disagreement between the involved agencies, contradicting the allegedly integrated approach:

> The Agency for Cooperative Promotion of the Amhara National Regional State has initiated the formation of the Koga Irrigation Cooperative. This is quite substantial. But, the articles referenced from the proclamation pertaining to the establishment of cooperatives are not in most cases suitable for the establishment of an irrigation management organization, namely an IWUA. This has been contentious between the Consultant on behalf of the Client [i.e., the MoWR] and the Agency and has been viewed by the latter as an encroachment into what is considered by them as justifiably the Agency's sphere of activity.[3]

Busy with organizational and institutional confusion, the ministries failed to take other, potentially more important issues of participation into account. This failing led to a situation in which the process of decision making within the farmers' organizations substantially reproduced social inequalities: those who already possessed power in the respective community also filled the most important positions in the irrigation cooperative.

LOCAL REPRODUCTION OF POWER RELATIONS IN IRRIGATION MANAGEMENT

The mechanism of reproducing power is rather simple according to both the leaders' perceptions of why they were voted in and the members' statements on why they voted for someone. According to interviewees, the most important characteristics a person had to have in order to be voted for were (in descending order of importance) literacy and experience dealing with government officials. The criterion of literacy reduces the number of possible candidates considerably, as about 80 percent of the rural population in Amhara are illiterate. It also makes the election of women to leadership positions even less likely considering the difference between male and female literacy (around 30 percent and 10 percent, respectively; see Shenkhut 2005). During research, no women were or had been active in any position of the cooperative.

Because basic literacy (as well as mathematical literacy) is crucial to fulfilling the tasks that come with the official positions in the organization, the reproduction of power along already established hierarchies makes perfect sense in a technocentric understanding of farmers' institutions. The problem is rather that the needed basic skills cannot be acquired by most.

The second point, namely, the capacities required to deal with government officials, especially applies to the higher positions in the organization and narrows the potential candidates to a small proportion of politically active people. Being familiar with handling administrative affairs and dealing with bureaucratic structures in the rural context usually comes with working for political parties or administration at the kebele, or peasant association, level (the smallest administrative unit of Ethiopia).

In this context, it is important to understand that the administrative institution of the kebele was established by the Derg regime in 1975 as a political instrument through which the regime "literally controlled every village and every human activity in the vast rural areas of Ethiopia" (Aadland 2002, 36). The kebele also played an important role in the prosecution of political enemies during the Red Terror campaigns. Although the leaders of the kebele were replaced after the downfall of the Derg, the structures were not, and the new ruling party could soon restore control through their own executives within the kebele structures (Pausewang 2002, 98). Over time, this newly exerted control from above increasingly resulted in a situation in which the "kebele are once again monitored and run by political cadres" (Aadland 2002, 36). Most of the IC's board members held such a position in the past or are still active in local party politics.

CONCLUSION

As the study has shown, food security in Ethiopia is, in many respects, a political problem. Earlier research on disasters and on famines, in the Horn of Africa in particular, suggests, as pointed out earlier, that vulnerability to droughts "is known to arise from political marginalization rather than either technical deficiencies or the vagaries of the weather. . . . In brief, the real issues underlying the persistence of famine are about the lack of political inclusion, not the lack of technical interventions" (Lautze and Maxwell 2007, 239 f.).

The multiscalar analysis revealed how policy narratives on the character of water resources management in general and irrigation in particular travel between the political scales. New policies and paradigms that are produced as an effect of changing global discourses have concrete impacts on power relations between actors at different scales.

The case study showed that global paradigms of how irrigation water is supposedly managed are best manifested on the local level through the intervention of the state. The "WUA discourse" is a good example. Farmers had to deal with the contradiction of being pushed to change farming practices for commercial production as a result of IWRM-related policies, although the necessary inputs for this were not available to them. These underlying reasons for farmers' "conservative" behavior went unnoticed in the ministerial debates. A closer look at the linkages between the different political domains reveals that while global politics and institutions constrain the agency of the state by imposing certain policies on it, they also enable government actors to pick and choose from available discourses.

Climate change legitimizes infrastructure development in the face of transboundary hydropolitics. The Ethiopian government can extend its scope of agency with reference to the rather new issue of climate change and the surrounding policies like the NAPAs. However, while the implementation of irrigation projects, like the one in Koga, might mitigate the severity of disastrous water-related events, it does not necessarily lead to a decreased *vulnerability* to floods and droughts on the local level. Current disaster research points out that marginal groups are more vulnerable to disruptions, while elites, both local and national, might even be able to strengthen their position. Thus any means taken to mitigate possible impacts of climate change and resultant extreme events have to effectively include those most vulnerable groups. Otherwise, existing inequalities within our "species" are likely to increase to the detriment of those who have contributed little to the sociogenic changes that the Anthropocene brings about.

NOTES

1. Publications reflecting such degradationist discourse include *The Population Bomb* by Ehrlich (1968), "The Tragedy of the Commons" by Hardin (1968), and *The Sinking Ark* by Myers et al. (1979).

2. NAPAs are a reporting process for Least Developed Countries to the United Nations Framework Convention on Climate Change (UNFCCC). These national reports are meant to identify priority activities that respond to immediate needs to adapt to climate change.

3. Unofficial working paper, "Irrigation Water Users' Association: Concept and Concern," by an MoWR training officer.

4

Rivers at the End of the End of Nature

Ethical Trajectories of the Anthropocene Grand Narrative

Celia Deane-Drummond

In considering the movement of the global to the local human scale and vice versa, I begin with two propositions. First, it is necessary to consider the ways in which humans tell stories or narratives about river systems. These stories impinge on the reasons for how they act; that is, they function in the sphere of morality and ethics.[1] Second, the ways that humans perceive their interaction with the natural world has shifted from being understood primarily as *makers* of technology to that of *consumers,* expressed most powerfully in the consumption of water and even water systems.

Given that stories—especially histories but also scientific narratives about the environment—operate in the realm of ethics and morality, it is important to understand that ethical frameworks, generally speaking, will have some idea of a goal of human flourishing. Who defines what this goal might be is crucial. In constructing their narratives, authors often embed their own assumptions about what is right or wrong—perhaps without consciously recognizing that it shapes their stories. These narratives can creep into scientific accounts, even those that use quantitative material. I am going to probe this further by looking specifically at the philosophical assumptions buried in the concept of the Anthropocene, which started out as a geological concept but has now moved beyond this into other areas of discourse.

Water has always been significant for human societies and for religious practices in particular. And access to water that river systems provide has shaped not just the historical development of ancient settlements but also the possible type of ecological interactions between humans and other species. The particular entanglement between humans and river systems is interesting, because it provides a case study for reflection on the way humans envisage their specific ethical responsibilities. But rather than focus on water as such in this chapter, I am going to address primarily the idea of the Anthropocene and its ethical impact, because the theme of this project, Rivers of the Anthropocene, weaves the two together.

The *Anthropocene,* if we use the definition coined by Paul Crutzen, refers to the geological epoch shaped by human activities since the early Industrial Revolution (Crutzen and Stoermer 2000). According to this narrative, for the first time in human history, humans have become such a dominant force that they determine the state of the earth's crust. So if we were to fast-forward in time, their imprint would appear in the geological record. The concept of reading history through the geological record is certainly not new; even the nineteenth-century geologist David Thomas Ansted (1863) viewed his work vividly as a reading of the "great stone book of nature." The pages may be crumpled or torn, or even inverted, but the book when reconstructed tells a tale of progression and change, with increasing significance as it approaches the geological era in which we live today. The sociologist and philosopher Bronislaw Szerszynski (2012), in his discussion of the cultural significance of the Anthropocene, argues that geology is, like medicine, *ideographic,* dealing with ideas that stem from particulars, rather than physics, which is *nomothetic* and deals with general laws. In order to convey such particularities in a meaningful way, geology makes particularities observed meaningful by a reading of signs; that is, it becomes *semiotic.*

But for the Anthropocene, the particular way in which one might read the sign in geology is complicated by the different historical phases of human activity and action. One image of humans in a technological world is that of *Homo faber,* humans as maker, in which the natural world is manipulated for *human purposes and ends.* In the Anthropocene this becomes superseded by *Homo consumens,* humanity the consumer, and *Homo colossus,* a term coined by William Catton that signifies humans as bent on consuming exhaustible resources, including water, leading to water scarcity. The number of references to such overconsumption of river systems at the Rivers of the Anthropocene conference in 2014 was striking (Kelly this vol.), but such shifts have a profound impact on human communities as well. So Szerszynski (2012, 175) suggests, "If the 'bad' Anthropocene has indeed been this parody of the cycles of nature—a growth without decay, a piling up of things which are at once consumed, a technological metabolism which turns nature into commodities without replenishing nature's self-reproductive powers—then it has been not the apotheosis but the eclipse of man as *Homo faber:* the end of the end of nature."

But the language of the Anthropocene has still more significance, because some scholars have started to argue that humans might be able *to replace* the destructive habits of *H. colossus* with a new approach that self-consciously manipulates the planet in ways that are viewed as positive and have good results. According to this approach, sometimes known as eco-modernism, any uncertainties are exogenous factors that can be dealt with through refinement of technique, a return of *H. faber* or even perhaps *H. melius,* one who makes things better than before.

But knowledge of the earth system and even detailed analysis of one river system within the total earth system already challenges such optimism as facile and far too simplistic, encouraging a replacement of mere humanity the maker, *H. faber* with *H. gubernans,* the helmsperson, steering the processes of the natural world in a particular direction. The close entanglement of humans with complex river systems is another reminder that bringing back the innocuous view of *H. faber* is naive. The nature of the river system itself and the living creatures within it will dictate what may be possible for humans in that particular system, and so on to a global scale. Twenty million inhabitants have been displaced in the Gulf of Mexico following the introduction of dams (Syvitski 2014). This case shows how attempts to steer the natural world to particular human desires, *Homo gubernans,* have gone awry and led to unforeseen consequences.

But I want to reflect further on what happens when we consider humanity through what could be called the gaze of the Anthropocene. Now humanity becomes woven *into* a geological system in a way that points to the fatalistic ending of human activity that then impinges on how humans act today. In this sense the dominant narrative of the Anthropocene is brutally consequentialist. The very process depends on the perceived consequences of that activity, even though its proponents usually claim that the language of the Anthropocene is ethically "neutral" or descriptive of "facts" and so somehow removed from a moral standpoint. The Anthropocene as a way of telling the human story thereby echoes more pessimistic discourse about climate change, which also conveys apocalyptic scenarios of humanity's demise.

The social scientists Andreas Malm and Alf Hornborg (2014) add another critical voice to presumptions embedded in the Anthropocene grand narrative by arguing that it jumps far too quickly from a natural science, geology, to an assessment of the impact of the whole species, *Homo sapiens,* thus missing the critical and crucial, textured, *social* elements that are woven into human histories. They also argue that the Anthropocene makes far too many presumptions about the activities of the human species as such, so that when dealing with climate change, a fossil fuel economy, for example, is certainly not attributable to the species at large but only a small fraction of that species. As such, it is misleading in its claims for the delegation of human responsibility to the whole of the species. This ignores inequalities in contributions by different societies to change. The Rivers of the Anthropocene project, by concentrating on local as well as global impacts, avoids this difficulty to some extent, but the problem is still evident, in that the term "Anthropocene" implies that the species as a whole contributes in some way evenly to disruption of water systems.

The Anthropocene is thus a geological grand narrative that carries cultural significance beyond its immediate scientific reference and coalesces with other grand narratives about climate change, pointing to the entangled fate of ecological systems, human beings, and river systems. The Anthropocene, with its specter of the ultimate end of humanity, can lead to very diverse ethical responses, from resignation through to revolutionary political action. These diverse ethical responses echo alternative narratives about the way humanity is connected with the natural world, either envisioning humanity as one species among many biota, such as in Lovelock's (1987) Gaia hypothesis, or humanity in a privileged position over and against the natural world, bolstered by the promise of new technologies. The difference between narrative and drama is important, because narrative rhetoric will, arguably, have different outcomes both ethically and politically from dramatic rhetoric (Deane-Drummond 2010).

What is meant by the term "narrative" and its particular function in religious terms is very diverse. I bring in religion at this juncture, as I believe that religion has a powerful influence in shaping both morality and ethics. In religious views, ethics more often than not take the form of a normative ethics, meaning what is right or not is laid down through given principles. Religious narratives can also reflect simply the nature of religious experience, so religion is about the way people tell particular stories, or how such stories give structure to the world and try to make sense of it. Narrative can mean more than this, however. It can also mean not just the form in which an encounter with the sacred takes place but also the bearer of the sacred. It can refer to the life story (biography) or experiences of a particular group or individual. Another form of narrative relates to the manner in which biblical text is set forth. Finally, portraying theological issues through narrative implies the use of narrative as a hermeneutic tool (Stroup 1984). I argue here that a greater emphasis on *drama* is important for ethics from a secular as well as a theological perspective when faced with the grand narrative of the Anthropocene. While I cannot do justice to the full ramifications of this attention to drama in the present context, I seek to give sufficient indicators in order to generate debate and discussion on this issue.

A traditional way of reading history is through genealogies or through a systemization of the dynamics of historical change in various sorts of grand narratives (Lovelock 1987, 2006; Swimme and Berry 1992). This is also true of "cosmic" history, including that expressed in various grand proposals, such as the cosmic creation story of Thomas Berry and Brian Swimme or the Gaia hypothesis of James Lovelock. For Lovelock the biota as a whole contribute in a vital way to the stability of the planet's life, such that the gaseous composition of the atmosphere and temperature are kept within the boundaries that are suitable for life as we know it. Human players are intimately bound up with this narrative inasmuch as only humans are conscious of what is happening, and for Lovelock this consciousness is in some sense *representative* for the whole earth (Deane-Drummond 2004). Yet, in spite of such an elevated awareness, as far as the earth is concerned human beings do not seem to contribute to its flourishing. One might even view humanity, according to some readings of the Gaia hypothesis, as a cancerous growth on the planet, bent on its destruction. The Anthropocene narrative is on a par with the Gaia hypothesis; it is a grand story about the fate of the earth in an era when humans now dominate. Like Gaia, it started off as a scientific theory, and like Gaia, it is becoming adopted more widely. The difference is that the Anthropocene seems less controversial when compared to a global Gaia hypothesis. In the Anthropocene the portrayal of the earth is more mechanistic, in contrast to the biological model from which Gaia takes its cue. In this respect I part company from Bruno Latour (2013), who, it seems to me, has conflated Gaia with the Anthropocene.[2]

Grand narratives create an aura of determinism, in which what is anticipated seems an almost inevitable trajectory of the story as told so far. For this reason, many historians eschew narrative as inherently teleological. As noted above, grand narratives level out social and cultural differences. The ethical and political implications of such epic readings are clear. If humanity is inevitably caught in a narrative that is of its own making but it is unable to change, forced into a new epoch that seems irreversible, then this will lead to fatalism. Politically, this will mean a shift in emphasis so that pessimism about any positive outcome of human intervention prevails; no action at all will be taken.

In practice, actual political activity is far more complex than this account suggests, in that confidence in the more traditional accounts of science on which the Anthropocene tends to rely overlie more subversive notions of science as represented in an expanded ecological account, such as in Gaian theory. In this case, fueled by dreams of a "good" Anthropocene, the competing narrative is more promethean and far more optimistic about humanity's ability to manage its own affairs, including the problems associated with climate change.[3] Mitigation and adaptation are viewed as both desirable and possible, such that collective human action is sufficient to counter any dangers or threats to human survival.

In the face of acute global water shortages, the Anthropocene may, indeed, encourage a revolutionary politics that assumes that human beings are inevitably caught up with social, political, and cultural change in such a way that weakens any sense of individual agency. This, ironically in the context of Western thought, amounts to a disassociation of the human as a political being from an understanding of humanity's embeddedness in the natural world, a way of perceiving political life as that constructed both by and for human agents. Perceiving the human in political terms as dissociated from nature has been a dominant strand in the Western history of the politics of nature. Peter Scott argues convincingly against such a view in favor of a postnatural politics. In his view, humanity's embeddedness in the natural world becomes not just an adjunct to human affairs, but a new approach to the political realm that weaves in a deeper understanding of human beings as part of nature, which is itself complex and constituted by unsettled boundaries such as that between the natural and artificial. For Scott (2011), "postnatural" does not mean so much *leaving behind* the natural as it does being aware and conscious of how deeply we are embedded in natural processes. The term is also intended to signify looser boundaries between the artificial and the natural, the human and the nonhuman, and thus challenges the notion of "natural" as a distinct category.

It is important to note that drama does not eclipse all narrative; rather, by giving attention to the lyric understood as individual and specific experience, it ensures that it does not slip into grand narrative or epic mode. Ancient literary criticism argued about the extent to which drama as a genre contained narrative, or whether drama imported narrative accounts as it were from the outside. Philodesmus, for example, argued against the position of Aristotle, who, in the *Poetics*, distinguished epic clearly from tragedy, calling the former narrative, or *apangelia*.[4] Yet the figure of the messenger (*angelia*) in Greek tragedy allows drama to express the narrative voice in a way that closely resembles an epic account. Hence the mix is present in the drama, but narrative is mediated through the messenger. But what precisely is the difference between epic narratives and drama?

Drama is most commonly represented as that which displays human actions and temporal events in specific social contexts. Drama reflects the indeterminacy typical of human life, including the unforeseeable interactions of circumstances and the ambiguities of existence (Balthasar 1988, 17). I suggest that focusing on a *local river system* and its specific instances of human-natural interactions can move the ethical discussion toward a dramatic approach. Rivers of the Anthropocene is, therefore, in a paradoxical way articulating *both* a narrative and a drama. It is a grand narrative of the global wedded to the specifics of the drama of the local. But I suggest that there are important religious and ethical consequences of each way of perceiving. Drama also has the characteristic of an "event" through the dynamic staging of particulars in a particular way. It also has an irreducibly *social* dimension, including the audience as much as those taking part in the play. In addition, drama includes the idea of anticipation, but this is not the same as resignation; rather, it is ongoing, consuming involvement in the work of interpretation (Quash 2005, 35–37). Drama certainly has the capacity to take up narrative elements, as is in evidence in classical Greek tragedy, but the difference from epic is striking, so that "Greek tragedy confronts the spectator directly with a multitude of voices, each with an equal claim, in principle, to truth and authority. The absence of a narrator renders all speech on stage equally authoritative or suspect, equally bound by its status as a rhetorical creation" (Barrett 2002: xvi). These encounters of different voices produce both opacity and ambiguity in language, and to some extent reproduce what occurred in the ancient political arena of Athens.[5]

The political implications of such a move toward the dramatic are, I believe, highly significant. In the place of resignation fostered by the Anthropocene grand narrative there is a greater stress on the importance of individual human agency, or at least agency in the context of a specified community. Yet because such agency is one that is caught up in a drama, it does not collapse into either individual liberalism or collective political liberalism but invites what might be called a version of postnatural politics, one where the human and nonhuman creatures are embedded and woven together in a drama. Collective and communitarian action is one, therefore, that is inclusive rather than exclusive of the nonhuman realm. Here one might envision river systems as being much more than simply a stage on which human activity is played out, since it is responsive to the multiple activities of all the different agents in the play. Yet I would press against the idea that the earth as a whole has agency in such a drama; rather, other players are all those creatures or perhaps specific elemental forces within the overall earth system that exist in active relationship with human beings and other agents. While humanity will be aware of its role in such a drama and animals will have a greater power of purposive motion compared to, for example, plants, much will be hidden from view, because the way the drama unfolds will not be known from its beginning, middle, or end.

Close attention to local issues at the human scale in river systems opens up a particular way of discerning that gives significance to individual human agency. The language of practical wisdom, or prudence, gives particular ethical insights into how humans might act in difficult situations in which there are conflicts of interests. It does this through drawing on the classic tradition of community discernment that entails a combined approach, including *memoria,* or memory of the past, that is authentic, *docilitas,* or teachability, circumspection, or taking account of concrete situations, including the science, insight, and foresight, as well as caution, that can be broadly related to the secular concept of the precautionary principle. Practical wisdom was developed in the ancient classical world through Aristotelian metaphysics. Aristotle's work was elaborated by Thomas Aquinas, who combined philosophical insights with a theological perspective that included the idea of divinely infused virtues that he drew from Augustine of Hippo. This ancient approach to practical wisdom, combining discernment, judgment, and action, resonates with the preference for the dramatic that I have indicated as important. Further, by including other players in the drama, the process of deliberation can include paying attention to other creatures and their desire for flourishing, as well as limiting a perspective based on a narrow definition of what the good might entail in purely humanistic terms.

How might prudence work when the problems associated with water governance are, in business parlance, sometimes thought of as "wicked"; that is, they lack consensus, are highly complex, and do not seem to be solvable (Camillus 2008: 99–106)? I suggest that while such complex problems may appear to be insoluble there is little excuse for inactivity, because attempts at a better resolution are ways of testing workable alternatives. So the exercise of practical wisdom is still open to risk, but it is a risk that has to be taken. Timothy Carter and Mary Miss hint at this when they suggest the importance of being proactive in the local context of urban ecology rather than simply reactive after a disaster (chap. 11, this vol.). Helen Berry also suggests that historical responses to the flooding of the Tyne River in 1771 bring to the surface important lessons in how to anticipate social unrest in times of natural disaster by looking closely at micro-histories—I would like to term these "dramas"—where exercise of prudential reasoning worked across political and social divides (chap. 9, this vol.). Further, once a religious dimension is introduced, as shown historically in Berry's account, the dynamics of the local drama shift to include religious belief in God, a theo-drama.[6] Religious dynamics, for good or ill, have shaped human entanglement with the natural world in terms of its history, productivity, and impact, including especially the entangled history of humans and riverine systems. As Philip Scarpino points out, the mounting evidence of unintended consequences of human activity are written into the earth's history (chap. 8, this vol.). These are haunting reminders of the human inability to act with prudence. Yet becoming aware of our place in the drama as participants and not mere observers is, perhaps, the first step.

ACKNOWLEDGMENTS

I am grateful for the invitation to join the Rivers of the Anthropocene project and to the conference participants for helpful discussion and feedback on material presented that formed the basis for this chapter.

NOTES

1. I define *ethics* as a reasoned study of *how* humans act in a given society. It is, at its core, *evaluative* about what is right or wrong. This distinguishes it from a field such as history, which considers *why* humans act historically—ostensibly avoiding any judgment of right or wrong. Related to ethics is the issue of morality. I am defining *morality* as the presupposed or the self-conscious understanding in a given culture of what is the right thing to do. Morality, in its turn, is shaped by the particular way humans tell stories or narratives. Thus there is a relationship between making narratives and the construction of morality within a given culture.

2. See especially Bruno Latour, Lecture 3, "The Puzzling Face of Secular Gaia," and Lecture 4, "Anthropocene and Destruction."

3. For a discussion of the ethical ramifications of very different portrayals of the Anthropocene as "good" and "bad," see in particular Antonaccio 2017.

4. Aristotle, *Poetics*, 144 9b11, cited in Halliwell 1986, 128–29.

5. The figure of the messenger allowed for implicit claims to a more secure form of knowledge, such that his report presented itself as an unproblematic and nonrhetorical account of events offstage. The figure also allowed for some narration of events that could not easily be presented in a play, such as miracles (de Jong 1991, 117).

6. I develop a theological anthropology that develops the significance of theo-drama for human action in Deane-Drummond 2014.

5

Rivers, Scholars, and Society

A Situation Analysis

Kenneth S. Lubinski and Martin Thoms

Scholars, the societies we live in, and the institutions responsible for river management need to accept, understand, value, and succeed at living within the limits of our natural resources. This need applies across cultures and political systems. Rivers, because of the services they provide to humans and other forms of life, are effective ecosystems for demonstrating the conflicts that arise when humans do not learn how to curb their desires or share the benefits of nature. Scholars, people with advanced knowledge of a subject (usually taught in school), play an important role in showing society the consequences of its decisions and actions. We must learn and effect change at a rate that is faster than the rate at which humans are currently using and degrading earth's limited resources. Rivers, then, are high-visibility test cases for evaluating whether scholars in particular are functioning effectively in society.

One premise of the Rivers of the Anthropocene project is that a transdisciplinary approach by scholars will be more effective than single-, multi-, or interdisciplinary approaches to helping societies manage rivers (Kelly, this vol.). The expectation is that historians, scientists, artists, economists, and anthropologists, to name some of the scholar tribes, can develop more relevant and instructional sets of evidence and merge them into more influential messages when we work in collaboration. Palsson et al. (2013) described the need to take the first, collaborative learning step as a way to "reorganize our house" in preparation for helping society halt or reverse the impending environmental crisis. The implication is that the role that scholars play in society, that of village elders or wise men and women, can no longer be played adequately if we perform only as individuals or cliques.

It is a noble premise, and one that is difficult to criticize. But it requires additional thinking about the desired endpoint and the practical issues of getting there. Is the intent only to learn and inform together activities that are mostly under our control and measurable using traditional academic metrics? Or is success to be measured by real increases in river ecosystem quality, slower rates of river degradation, or wiser and fairer allocations of freshwater? If the latter, we need to identify the publics, institutions, and stakeholders we want to influence and develop common strategies to exert that influence. Success in this case will not just be under the control of scholars, but the result of scholars interacting effectively in diverse societies, cultures, and communities.

This chapter considers scholar-society relationships based on observations of past circumstances and likely future interactions. The relationships are complex, and thus they present challenges to the concept of traveling a more transdisciplinary path and arriving at the intended destination. The observations are framed here as a situation analysis—addressing where we are in time and establishing a base from which optional paths forward can be considered. "We" here usually refers to the broad community of scholars—natural scientists, social scientists, and humanists (Palsson et al. 2013), as well as economists, all of whom have specialized knowledge and make their living by learning and teaching, regardless of discipline, institution, or audience.

Our intent is to start at the end of the proposed journey and work backward. We begin by clarifying what we think is the collective desired endpoint: a future in which rivers are managed sustainably, in ways that adequately address the interests and requirements of diverse stakeholders and nonhuman species. From there we discuss how scholars need to function effectively in the societies that will have to accept responsibility for sustainable management. Last, we address scale, a special challenge of river socioecosystems and their future management.

A POTENTIAL DESIRED FUTURE: SUSTAINABLE MANAGEMENT OF RIVERS AS SOCIOECOSYSTEMS

For anyone, let alone two river scientists, to postulate a long-term societal goal may seem a bit pretentious. But it is necessary to clarify why scholars are considering the need to develop a transdisciplinary approach to—what exactly? So let's accept, for the purposes of discussion, that a useful and relevant societal goal is to be able to manage rivers sustainably. The most common and general definition of sustainability, the one that implies our intent to leave future generations with as many, or more, choices as we now have, works as a reasonable starting point here.

But for sustainability to function as an operational goal in real river policy development and management, this definition requires elaboration, including details of *how* society should achieve it. There is, for example, the now-widespread belief that sustainability can only be attained if humans are accepted as active

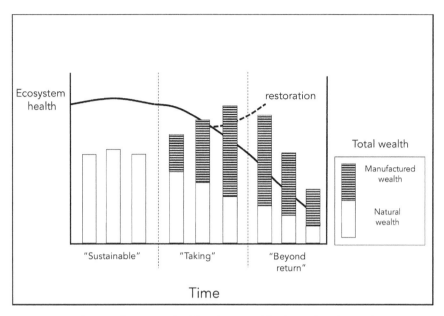

FIGURE 5.1. Theoretical ecosystem health and area wealth relationships during three stages of river use. Here "total wealth" includes both natural (equivalent to the natural capital of Daily [2003] and Karieva et al. [2011]) and manufactured wealth. During the initial stage, before humans became major environmental drivers, human use of a river results in minimal changes to the river's other ecological functions, such as its ability to support animal and plant species or biogeochemical processes. As the human footprint grows during the second "taking" stage, decreases in natural wealth are more than offset by increased manufactured wealth. The resulting increase in total wealth and benefits to humans are considered acceptable or even preferred over initial conditions, in spite of observable losses to other ecosystem functions. In the third stage, natural wealth and manufactured wealth both decline. The level of ecosystem health falls below a desirable level, and humans begin to see the river as a place to avoid. As the system moves from Stage 2 to Stage 3, more ecosystem functions are lost and a system state change occurs, from which restoration becomes virtually impossible in the foreseeable future.

components of ecosystems rather than as external driving factors. The inclusion of humans in ecosystems was one of the most substantial recent changes in natural resources management philosophy, a shift referred to as ecosystem management (Grumbine 1994). Adaptive management (Holling 1978), which recognizes that goal setting for complex systems is uncertain and that the iterative blending of learning and action is vital, is becoming more common where it is practical and realistic (Lee 1999). So progress has already been made. Many scholars are also exploring not only how ecosystems, cultures, and economies are linked, but the notion that "strong" sustainability requires the acceptance that economies are directly and strongly *dependent on* natural resources (Walker 2012). Figure 5.1 captures one interpretation of how river ecosystem health (Lubinski 2010), natural wealth, and manufactured wealth interact with each other over a long period (the Anthropocene?) of increasing human use. If, collectively, scholars can quantify such relationships, the resultant knowledge could be applied to rivers before they degrade past the point of potential restoration.

Together, the above beliefs, among others, have given rise to the idea of treating rivers as socioecosystems (Machlis, Force, and Burch 1997; Folke 2006), which has great promise as an organizing concept. By evaluating rivers as socioecosystems, we are forced to accept trade-offs and identify minimally acceptable standards for all of the system components of interest. Multiple use dictates that no single user can have it all (Cairns 1972). But putting the concept of rivers as socioeconomic systems into practice will introduce many new policy and management hurdles. Presented as questions, some of these are:

· What institutions are ready, willing, and able to operate in this way? Do the existing institutions have adequate authorities to implement socioecosystem policies?
· What decision-making processes are appropriate for evaluating trade-offs, and who will be given the responsibility of making such trade-offs in an unbiased, fair, and transparent way?
· What models are suitable for adequately describing river socioecosystem complexity and uncertainties?
· How can boundaries be drawn around a river socioecosystem in a way that internalizes all of the relevant parts and relationships?
· How should governance processes be modified to promote the holistic concept of river socioecosystems but also to give voice to their diverse stakeholder groups that need to take part in the functioning of these systems?
· How will the linear, long (often interjurisdictional), and integrative nature of rivers make their treatment as socioecosystems even more difficult?

These will not be easy questions to answer. And scholars will not be the only people responsible for answering them. Scholars will act in the role of consultants. Not all scholars will engage, but those who do will need to start tailoring their plans to address these and related questions.

SCHOLAR-SOCIETAL RELATIONSHIPS IN THE DESIRED FUTURE

The questions listed above are probably not too hard for policy makers and managers to anticipate. The larger question for scholars, however, is what do we have to do, collectively, to keep the goal of sustainable river socioecosystems realistic, attractive, and feasible in the eyes of the public and decision makers. The changes that will be required in the way river institutions and publics think and act will take time and energy. Scholars can play a key role in activating the transition.

The existing institutions are invested in the status quo, and most are not currently capable of changing their own responsibilities and authorities. In the United States, there are no singular institutions that have the responsibility or authority to manage rivers as socioecosystems yet. Therefore, the work will need to be accomplished through partnerships. Partnerships have become much more common over the past three decades, but, except in urgent situations, they have required more time to make decisions and take actions than single organizations. They tend to be politically popular for limited periods, especially when created to resolve a specific problem. But once the problem is resolved (politically if not actually), funding lapses.

The transition is likely to be led by policy makers, via effective communication strategies with river publics. We need to know the latter's beliefs, goals, and value systems. Do they listen more with their heads or their hearts? What cultural norms are in play? Excellent ideas are being explored by our colleagues on the subject of how fit existing institutions currently are to achieve goals related to sustainability (Costanza et al. 2001; Farrell and Thiel 2013).

Once scholars know the institutions well, we will need to develop well thought out strategies for effecting change. Clearly, change can be effected by more than just information. If we accept the model that human actions are driven primarily by their needs and beliefs, we can start asking who among us is best equipped and thus has a better chance to succeed along different causal pathways (fig. 5.2). For example, scientists and economists can provide the necessary evidence to convince managers that the effort is doable and relevant, while artists and historians may focus more directly on public beliefs. All of this work will require improvements in the way that scholars communicate with nonscholars. Facilitation skills will be vital. Members of policy-making groups should be invited to join us as early as possible.

Change is not something that only others need to implement. We scholars will need to take a good look at ourselves and ask what our strengths and weaknesses are relative to performing as a team of village elders. Working together in a transdisciplinary way to learn more deeply about river socioecosystems will require more than occasional communications and the publication of single-author papers in professional journals. Sacrifices of individual time and desires are always necessary for the success of a group, and these are perhaps the primary reasons that transdisciplinary approaches have not been common among scholars in the past. We will have to learn to play by a new set of rules intended to ensure team success, sometimes at the apparent expense of individual success and sometimes when success itself is not only dependent on how well we do.

We will have to become much more aware of what drives each other. The Rivers of the Anthropocene Conference and Workshop suggested that our individual reasons for beginning this discussion were as diverse as our disciplines. Fundamental differences in our perceptions and beliefs were hinted at, especially when we

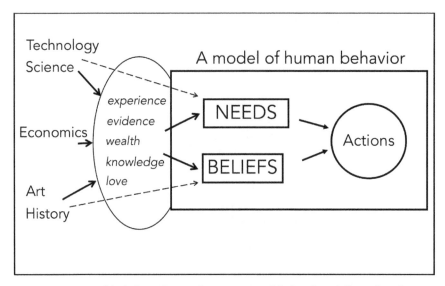

FIGURE 5.2. A model of what influences human actions. Scholars from different disciplines influence society (human actions) in different ways. Some disciplines, directly (dotted lines) or through their products (in italics), are more effective at changing human belief systems. Others are better at serving (and subsequently changing) important human needs. Collaboration among disciplines will require thoughtful partitioning of responsibilities to make adequate progress along both paths. Modified from Moore 1999.

discovered words and phrases whose definitions we as individuals were taking for granted. Figure 5.3, for example, initiated a debate about whether the terms "ecosystem integrity," "pristine," "natural," and "restoration" were still relevant to the management of rivers as socioecosystems. Our ability to persuade river policy makers and managers will be dependent on how we communicate as well as what we communicate. Agreement on where and when such terms should be used will be critical. An accepted glossary for moving forward will be an important task. But more important will be extended discussions explicitly intended to determine the extent to which we all truly want the same thing. Close inspection and spending more time together are likely to reveal commonalities and those concepts on which we have divergent opinions.

SCALE AS A SPECIAL CHALLENGE OF RIVER SOCIOECOSYSTEMS

Rivers are extraordinarily functional and provide a wide range of services to humans. It is in part their functionality that has led to the extremes to which they have been altered to serve even more human needs. The effort proposed in the

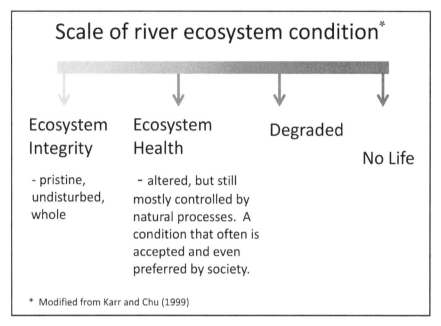

FIGURE 5.3. Noteworthy conceptual markers along a spectrum ecosystem condition. A simple attempt to clarify levels of ecosystem condition illustrates how disciplinary attitudes shape perceptions of terms and their values. This figure, for example, revealed deep differences in comfort levels among conference participants for the use of terms like "ecosystem integrity," "pristine," and "natural." The inability of scholars from different disciplines to agree on the definitions and values of such terms is a major challenge to functioning in a transdisciplinary approach. More important, this inability makes it extremely difficult for scholars to gain the respect and trust of societies that need to use such terms in emerging programs.

Rivers of the Anthropocene project suggests that we think of rivers as systems whose behaviors can be better understood if compared to each other. Many such efforts have been completed in the past (Oglesby, Carlson, and McCann 1972; Coates 2013). The exercise should expose, across many rivers in diverse cultures and political systems, common features of human-river relationships and foster greater understanding of the essence of those relationships.

But the concept of river socioecosystems has never been investigated at the level of comprehensiveness suggested at the conference. Ironically, a distinct feature of rivers makes them especially difficult for exploring the joint concepts of transdisciplinarity and socioecosystem management.

The functions and services that rivers provide cover a diversity of spatial scales. There are hierarchical networks of river basins and drainage networks—parts of the global hydrologic cycle that cross all manner of landscapes, ecosystems, and political and demographic boundaries. Basin landscapes are well-known, major drivers of river flows, water quality, and human use, but groups of humans that band together as river caretakers or managers seem to lose, except in special cases, connections with rivers and their basins at very large spatial scales. Hannon (1994) attributed this to the tendency of humans to discount the value of things that exist or happen at greater distances in space and time.

Many authors have described breakpoints in the hierarchy of river networks using relatively common terms such as "basin," "river," "reach" (usually between tributaries along a main-stem), "segment," and "habitat." Human interests in rivers can follow the borders of any of these levels of scale. On the Upper Mississippi River, as an example, human (community) perceptions of the river as a neighborhood seem to exist mostly at the reach level. A several-hundred-mile reach of this river, which mostly drains forest or dairy landscapes, is characterized by a narrow floodplain, and because of its fishing and hunting values was designated a national wildlife and fish refuge in the 1920s. Below this reach, however, the river's tributaries begin to drain flatter landscapes, which are dominated largely by agriculture and which now carry higher loads of nutrients and sediments to the main-stem. The main-stem floodplains, in turn, have been leveed, in part because upstream changes in the ability of the landscape has made downstream flooding more severe and less predictable. In this reach, the main-stem river begins losing its aesthetic appeal. Although it retains its value as a fish and wildlife resource, that value is held by a smaller percentage of the reach's public, the majority of whom see the river as something that needs to be kept on the other side of the levee.

Most large rivers are characterized, like the Upper Mississippi, by a small number of distinctive reaches defined by hydrologic or physiographic features. But histories of human usage, highlighted by dams, diversions, floodplain land use, and levee systems, have also provided artificial boundaries that exert powerful influence over the perceptions of nearby human communities. River system differences may well outweigh their commonalities from the perspective of determining important community boundaries or managing harmoniously across scales.

CLOSING POINTS

Scholars, in addition to their teaching duties, have often played the role of village elders in society, but usually that role has been carried out on an individual basis. When scholars have worked together, much of that work has been within professional societies, organized along disciplinary lines as opposed to societal problems. Working collaboratively to help society learn and act to achieve a very complex goal that has yet to be commonly valued and accepted will be challenging in predictable and unanticipated ways.

It may not be necessary to create a single common scholar's vision in support of sustainable river management. Such a vision would by necessity be so prescriptive as to erode the diversity and creativity of thought that scholars value so highly.

But success will require sacrifices of individual control and rewards. Academic institutions will need to adapt by recognizing the value of such collaborations and making the necessary resources available.

River systems are iconic in terms of representing human–natural resource relationships that will be vital in achieving future sustainability. We must learn and effect change at a rate that is faster than the rate at which humans have and are currently using and degrading earth's limited resources. Society does need to realize that the traditional approach, taking care of the economy and human conveniences first and fixing the environmental problems later, is not going to be viable in the future. If scholars can learn how to be effective participants in the management of these highly used socioecosystems, that learning can be applied to virtually any other system at less risk of overexploitation.

PART TWO

Histories

Everything has a history, and so it is with the concept of the Anthropocene. Good history sharpens our view of the past, connects past to present, and provides guidance as we look ahead. History is also constructed in the present; it is as much a product of the time in which it is written as the period it seeks to analyze and explain. Each of the chapters in this part (and all of those in the book) emerged directly out of the present-day knowledge and perspective of scientists and other professionals who examined past human engagement with earth systems, crafted an explanation that is compelling in the present, and presented a framework for understanding and informing the relationship between people and their environment as we advance into the future. History here takes on additional complexity, ranging from "deep" Holocene-era anthropogenic change to more recent scholarly investigation of human-environment interaction.

It has been over seventeen years since Paul Crutzen declared in 2000 that the earth had entered a new geological epoch in which human action had become the driving force in shaping global environmental change—human action so profound that it left a sedimentary and eventually a stratigraphic record. In chapter 6, Zalasiewicz, Williams, and Smith describe the Anthropocene process as recording "a significant geomorphic signature," which "will continue to evolve and leave a distinctive stratigraphic signature as the cumulative effects of anthropogenic changes work through the earth system." The Anthropocene removes what has become an increasingly artificial divide between human and natural history. As Scarpino notes in chapter 8, "Through the lens of the Anthropocene, the boundaries between natural and human history blur; understanding the present-day environment requires paying as much attention to human agency over time as it does to the evolutionary trajectory of natural processes."

The authors in part 2 consider the Anthropocene through multiple historical lenses—from the geological history of the English Fenland to the industrial history of the Seine to the history of environmentalism. Zalasiewicz, Williams, and Smith offer a case study (with much wider application) of the creation of an Anthropocene landscape in the Fenland, located on the east-central coast of England. They consider a deep history of Holocene deposition extending back to 7,690 B.P. Setting it in the analytic context of profound anthropogenic change, they link contemporary narratives about the past to the geological revolution of the nineteenth century—in their example, linking their research on the Fenland to Sydney Skertchley's geological memoir published in 1877. This geological history points to the new futures for the Fenlands as a result of climate change. Meybeck and Lestel focus in chapter 7 on the River Seine, 1870–2010, from its headwaters to its estuary, noting that "river basins can be used to validate the Anthropocene concept: they are a key component of earth system analysis." They employ archival sources, archaeological investigation, and sedimentary analysis to develop a profile that encompasses reversible and irreversible changes of a much-altered river, facts that must be taken into account when developing management strategies for the Seine basin. Scarpino provides a historical context for the trajectory of scientific investigation and global environmental change that helps to explain the genesis of the Anthropocene and the speed with which the idea caught on once proclaimed by Crutzen. Drawing together the important continuum of past, present, and future, he notes that gaining useful insight "into what people did in the past, how they act in the present, and what they are likely to do in the future" requires paying careful attention "to the complex and subtle tapestry of culture over time."

6

An Anthropocene Landscape

Drainage Transformed in the English Fenland

Jan A. Zalasiewicz, Mark Williams, and Dinah M. Smith

Alterations to the global fluvial system associated with the onset of the Anthropocene have been profound (Syvitski et al. 2005; Syvitski and Kettner 2011; Merritts et al. 2011; Williams et al. 2014). They have involved both direct reengineering of river systems to, for instance, "stabilize" channels, prevent active meandering, and impound water in dams; and indirect changes resulting from land use change, commonly involving such phenomena as increased sediment supply from deforestation and urbanization. However, the spectrum of changes goes beyond such well-documented effects to produce some novel and geologically counterintuitive phenomena that have already produced a significant geomorphic signature. These phenomena will continue to evolve and leave a distinctive stratigraphic signature as the cumulative effects of anthropogenic changes work through the Earth System. Here we describe one such example, from the Holocene deposits of the English Fenland, in which an extensive buried channel system is spectacularly exhumed and then topographically inverted by regional anthropogenic modification. This geologically novel transformation will be a strong influence on the course of future change in the region as global climate warms.

GEOLOGICAL FRAMEWORK

The English Fenland covers areas of Lincolnshire, Cambridgeshire, northern Norfolk, and parts of Suffolk and is the largest area of Holocene deposits (some 4,000 km^2) in Britain. Fenland sedimentary deposits are up to 30 m (more typically up to 20 m) thick, and they show evidence of a complex paleoenvironmental history.

They exist atop a pre-Holocene surface mostly composed of Jurassic clays (French 2003) overlain by Pleistocene tills, sands, and gravels (Wyatt 1984). More resistant Chalk underlies the eastern and southeastern part of the Fenland Basin, and limestone occurs to the north and west. The paleosurface, on which the Holocene deposits rest, is uneven, and areas of higher altitude formed "islands" such as Ely, March, and Thorney. These "islands" are in effect inliers of older strata surrounded by Holocene deposits and are overlain by Pleistocene gravels and till (Hall 1996).

The Holocene of the Fenland has a long history of study (e.g., Skertchly 1877; Godwin 1978; Horton 1989; Waller 1994; Smith et al. 2012), made all the more remarkable because the geology—via rapid wastage of the peat and exposure of the underlying geology—was changing rapidly. So, as these successive studies took place, each generation of researchers was analyzing what was essentially a different landscape. The geological memoir of Sydney Skertchley of 1877 is a largely forgotten classic (Skertchley is now better remembered in Australia, where he later emigrated, than in England), in which close observation of the Holocene deposits is allied with sophisticated study of the tidal dynamics of the Fenland rivers, in a process-based approach that only became commonplace in sedimentary geology a century or so later.

The Fenland succession essentially comprises a tripartite succession of Basal (formerly Lower) Peat overlain by a thick clay-dominated unit (now termed the Barroway Drove Beds), in turn overlain by an Upper Peat (Nordelph Peat); a subsequent, fourth, stratigraphic unit, the silty Terrington Beds, has a more limited distribution to the north and east (fig. 6.1). The succession spans much of the Holocene, commencing an estimated ~7690 B.P. ranging to ~2250 B.P. for the bulk of the succession (Smith et al. 2010 and references therein), though sedimentation continued locally into Roman times and later, while accumulation of peat continued as peat bogs, locally raised, until this was halted, and then reversed by wastage, as large-scale drainage schemes came into operation in the seventeenth century (see below). Sedimentary accumulation today mostly takes place seaward of the seawall, in a relatively narrow prism of intertidal deposits.

ANTHROPOGENIC CHANGE AND REVEALED GEOLOGY

There has since been major change to this succession. The Fenlands were locally drained during Roman and medieval times, but thorough transformation began after the phase of seventeenth-century drainage associated with the Dutch engineer Cornelius Vermuyden (1595–1677), which has continued to the present day. The Upper Peat has almost completely disappeared through drainage and subsequent ablation ("Fen blows") and oxidation, together with some peat cutting for fuel. This was a unit that originally exceeded 4 m in places as seen from evidence such as Holme Post (fig. 6.2)—an iron post hammered into the ground with its

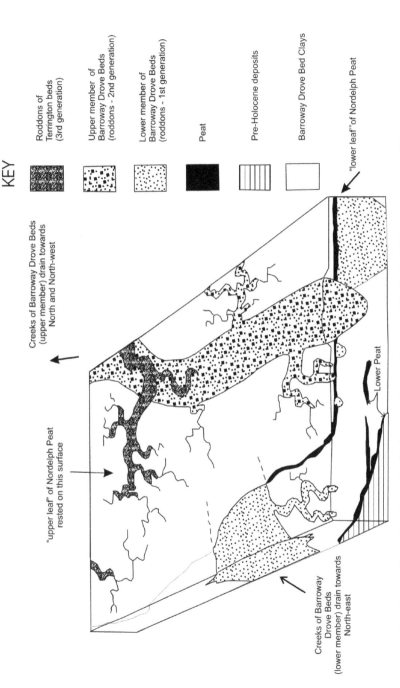

KEY

Roddons of
Terrington beds
(3rd generation)

Upper member of
Barroway Drove Beds
(roddons - 2nd generation)

Lower member of
Barroway Drove Beds
(roddons - 1st generation)

Peat

Pre-Holocene deposits

Barroway Drove Bed Clays

"lower leaf" of Nordelph Peat

Creeks of Barroway Drove Beds
(upper member) drain towards
North and North-west

"upper leaf" of Nordelph Peat
rested on this surface

Lower Peat

Creeks of Barroway
Drove Beds
(lower member) drain towards
North-east

FIGURE 6.1. Diagram illustrating the relationship between the main elements of the Fenland Holocene succession (after Zalasiewicz, in Horton 1989).

FIGURE 6.2. Left: Holme Post, Cambridgeshire (U.K. National Grid Reference [NGR]: TL 205895), showing previous ground levels (image courtesy of Hilary Welch, Conservation Projects for the Fens Tourism Group). Right: J. A. Zalasiewicz standing beside Holme Post in 2008, at about 2 m below sea level.

top at the peat surface in 1848 (i.e., well after drainage of the fens had begun), and the peat likely contained (at least) several hundred million tons of carbon, now released to the atmosphere (R. Eihenbaums, unpublished).

The underlying Barroway Drove Beds clay has been strongly compacted in its uppermost ~2 m, as it is dissected by a closely spaced, regular network of drains and dykes. From these, the water is continually pumped into the Fenland rivers, which are raised by a few meters above the surrounding landscape (which commonly is >2 m below sea level, with the sea being held back by an earth wall), through which it then flows out to the sea.

The landscape has thus been transformed in a manner without precedent in geological history. However, there are some near-parallels in peatlands around the world today. See, for example, the website of the International Peat Society (http://www.peatsociety.org/peatlands-and-peat/global-peat-resources-country) and estimates of peat loss in coastal settings such as the Netherlands (Erkens et al. 2016) and the Florida Everglades (Hohner and Dreschel 2015), many of which, in one way or another, have been profoundly affected by anthropogenic change. For the contemporary Fenland to fulfill its modern use (it is one of the most productive agricultural areas in Britain) it needs continual maintenance and pumping, akin to a patient on a life-support machine. How long that machine may continue to function with global warming is questionable (see below). Nevertheless, one of the results of Fenland transformation has been the exhumation of many major archaeological structures, formerly buried in the peat (e.g., Malim 2005; Pryor 2005), and of a remarkable suite of finely preserved channel structures.

THE FENLAND RODDONS

Incised in the top surface of the drained, compacted Barroway Drove Beds Clay (generally interpreted as former salt marsh clays) there are clearly distinct silt-filled drainage systems—dominantly tidal creeks that include a component of paleorivers—the remarkable, preserved remains of which are locally known as roddons (fig. 6.3). They have been recognized since the peat cover of the fens began to waste away (Skertchly 1877; Darby 1983; Hall 1987). In all, three generations (separate networks) of roddons have been identified (Horton 1989). Two extend across the Fenland, with markedly different channel patterns and orientations (Horton, 1989); a third generation is present mainly to seaward around the Wash (figs. 6.1, 6.4).

Each generation of roddons forms a hierarchical network, the main "trunk" roddons ranging from a few hundred meters to a kilometer across and traversing the entire Fenland area. These major Fenland roddons likely had substantial freshwater input, and a few have been identified as former courses of some of the extant rivers of the Fenland, such as the Ouse and the Little Ouse (Astbury 1958) and the Nene (Smith et al. 2012) Rivers of Cambridgeshire. Tributary roddons of at least

FIGURE 6.3. Roddons visible in fields as slight undulations at Plash Drove, near Guyhirn, Cambridgeshire (NGR TF385063). Optimum times for viewing roddons are during intervals when fields are crop-free (view looking north).

two more generations branch off from the trunk roddons, most of these smaller structures having blind ends inland and previously evidently were salt marsh creeks, with both water and their infilling sediment sourced directly from the sea. The smallest of these tributary structures are as narrow as ~2 meters across. The preserved depth (i.e., thickness of sediment infill) ranges from ~1 m in the smallest channels to in excess of 10 m in the trunk roddons.

The form, structure, and genesis of the Fenland roddons has been most recently examined by Smith et al. (2010, 2012). These structures contrast strongly with most tidal creek/meandering channel deposits in the geological record in that their form reflects preservation of a single channel thread rather than the sheet of laterally stacked point bars, reflecting successive phases in active meandering, which is the more typical record of long-lived meandering rivers or tidal creeks. This pattern strongly suggests that the roddons underwent a short-lived history of incision and then geologically instantaneous infill with silt and fine sand, an inference supported by detailed sedimentary analysis of spring-neap tidal cycles preserved within the infill (Smith et al. 2012), which suggest that the roddons may have converted from active channels to being more or less completely sediment-choked in

FIGURE 6.4. LiDAR image of the roddons in the Boston-Fishtoft and coastal area in Lincolnshire (at NGR TF329437). These roddons show up as dendritic patterns traversing the countryside from the coast. Roddon systems of the second and third generations (the latter is the youngest) are visible. These tidal channels, which ultimately silted up, are now visible on the LiDAR as roddons. Image courtesy of UK Environment Agency.

only a few years. Locally, a longer-lived depression filled with still or slow-moving water was left, which acted as a site for subsequent human habitation, leaving a rich archeological record in the more slowly accumulated final infill (Smith et al. 2012 and references therein).

We know of no exact counterparts of the Fenland roddon systems elsewhere in the world. Their characteristic morphology clearly reflects the particular context of the extensive nature of the premodified Fenland wetland landscape that was adjacent to a highly active tidal coastline. We have inferred that this coastal zone was subject to abrupt geomorphic change (e.g., through major storms) with consequent effects on the hydraulic geometry of tidal transport paths (Smith et al. 2010). Such changes to the coastal zone could plausibly have led to a change from the system being ebb-dominant (tending to keep the channel system scoured clean of sediment) to flood-dominant (tending to rapidly fill the channel system with sea-derived sediment)—hence rapidly producing the characteristic silty/sandy channel fills that are preserved as the roddons today.

Further, given that both of the main roddon systems occur at the boundary between the salt marsh clay deposits of the Barroway Drove Beds and the overlying peat, it may reasonably be speculated (Smith et al. 2010) that it was the choking of the tidal creek system over a wide area that restricted the access of tidal waters to broad areas of the Fenland, and hence led to the change from clay to freshwater peat deposition. This runs counter to interpretations from Skertchley (1877) on (e.g., Shennan and Horton 2002) that the clay-peat transitions represent sea level changes but is consistent with a recent global analysis (Lambeck et al. 2014) that sea level has been effectively consistent over the past six millennia, prior to its warming-related rise over the past century.

In the context of the present study, though, the significance of the roddons, which were formed as sediment bodies in the conditions of the Holocene, lies in their current morphology, revealed through differential compaction following draining, in what is now an Anthropocene landscape. Rather than being exhumed channel forms, they show strongly inverted topography, as sinuous ridges and (for the wider trunk segments) plateaus that stand up to 2 m higher than the surrounding clay-underlain surface. On the flat landscape of the Fenland, they constitute (other than the raised rivers and some "islands"—inliers of older geology) the only higher ground, such that the local farmhouses and other larger constructions are almost invariably built on them.

FUTURE EVOLUTION

The 5th Intergovernmental Panel on Climate Change report (IPCC 2013) predicts that sea level rise will be 52 to 98 cm by 2100 (and 26 to 55 cm even with aggressive CO_2 emissions reductions). These are conservative estimates, and may need revisiting following recent reassessment of twentieth-century sea level rise that indicates a more rapid increase over the past two decades (now at ~3 mm/year) than had previously been estimated (Hay et al. 2015). Regardless of the precise current trajectory of sea level rise, the formerly peat-covered Fenland area is now about 2 m below sea level (Ordnance Datum [OD]), with the silt-dominated areas lying at or just above 0.3 m OD (Waller 1994).

Over the coming decades and centuries, therefore, the Fenland is likely to be subjected to marine transgressions beyond the norm for the Holocene, and these will take place over an extensive area that has already been anthropogenically modified. The subsidence caused by the drainage (compaction) and peat wastage (removal of surface sediment) is effectively irreversible. It is clear that roddons cannot be reused as channels in future transgression events but will (together with the modern raised river structures) concentrate water flow between them. For a brief interval, before they too are submerged, they may provide a walkway system across the flooded landscape. The future geological record of the Fenland will thus include a striking Anthropocene signature, the result of human-driven modification of some remarkable channelized systems that had their genesis in a vanished sedimentary environment of the Holocene world.

ACKNOWLEDGMENTS

We thank Jason M. Kelly for the invitation to the Rivers of the Anthropocene Conference at Indianapolis in 2014 that stimulated this study and James Syvitski for constructive comments on the manuscript.

7

A Western European River in the Anthropocene

The Seine, 1870–2010

Michel Meybeck and Laurence Lestel

When Paul Crutzen, an atmospheric chemist, coined the term "Anthropocene" (Crutzen and Stoermer 2000), he was referring to a period when human control of the earth system, at the global scale, became equivalent to natural forces. River basins can be used to validate the Anthropocene concept: they are a key component of earth system analysis (Garrels, Mackenzie, and Hunt 1975; Berner and Berner 1996), providing information on regulating processes of the surficial earth (climatology, hydrology, production of vegetation, erosion, and weathering) and on the fluxes of material, water, nutrients, sediments, and so on, from continents to oceans. The Anthropocene concept was rapidly adopted within the scientific community; for example, the International Geosphere and Biosphere Programme (IGBP) has used it to describe rivers across multiple scales, from the local to the global (e.g., the biogeochemical cycles of carbon, nitrogen, and phosphorous; sediment fluxes and coastal morphology; water systems at the scale of continents) (Meybeck 2002, 2003; Vörösmarty and Meybeck 2004; Vörosmarty, Maybeck, and Pastore 2010; Seitzinger et al. 2005; Syvitski and Kettner 2011).

Another vision of river basins has been developed by environmental historians and geographers. It focuses on the multiple relations between rivers (and more generally water resources) and the development of societies since the Neolithic period. River basins have been essential to the development of agriculture, transportation, communication, food and fiber resources, and security. For this, rivers have been tamed, used, regulated, transformed, and sometimes diverted from one basin to another. In the pioneering book, *The Earth as Transformed by Human Action* (Turner et al. 1990), river basins such as those of the Thames and Nile were selected to illustrate longue durée interrelations (> 100 years) between humans and their environment. Other studies have focused on river uses and transformations over the past two hundred years (Mauch and Zeller 2008; Castonguay and Evenden 2012) or on the contemporary period (Arnaud-Fassetta, Masson, and Reynard 2013).

River-related activities can be reconstructed using data from historical and/or archeological archives as well as from the sedimentary archives in floodplains, deltas, and estuaries. This work has been done for a number of systems, including the Chesapeake Bay (Cooper and Brush, 1993) the Seine, the Spree, the Po, and the Zenne (Lestel and Carré 2017). In recent decades, environmental concern about rivers and their "quality" (i.e., their capacities to fulfill society uses in addition to our contemporary vision of what a "good ecological state" should be) has been developing. This has led to the Water Framework Directive (WFD 2000), which aims to restore all European Union (EU) rivers. The criteria for what constitutes a "good ecological state" are determined by societies, and they have evolved over the past 160 years. Well-documented river systems for which we have data on earth system processes (e.g., soil erosion, primary production) as well as social processes (e.g., population dynamics, ways of living, security needs) are particularly convenient for studying the evolution of the relationship between a river and society over the longue durée.

The Seine River basin (France), 65,000 km^2, fulfills these criteria: the Piren-Seine program, started in 1989, is studying the present-day functioning of the river basin, particularly the river quality, from its headwaters to the estuary, and its evolution over 140 years under changing demography, economic activities, water institutions, water quality regulations, and water sanitation. The program is highlighting the enormous influence of the megacity of Paris on the Seine basin and the major physical and chemical transformations that have greatly evolved over time (Meybeck, de Marsily, and Fustec 1998; Barles 1999; Garnier and Mouchel 1999; Barles and Mouchel 2006; Billen et al. 2007; Meybeck et al. 2007; Meybeck et al. 2016; Lestel and Carré 2017).

We consider first the distribution of the maximum physical and chemical impacts in the river basin at selected periods, first by stream order, a hydrological concept, second by the upstream/downstream impacts of Paris megacity. Then we analyze the general mass-balance of nutrients and material flow of metals in the basin, in comparison to their natural circulation rates in "pristine"—or preindustrial—conditions. The Seine longue durée analysis (1870–2010) shows large-scale trajectories and reveals both reversible and irreversible alterations of the basin. Finally, we propose a general scheme showing the stages of societal response to Anthropocene river basins, highlighting the remaining irreversible changes of basins, their regulations by societies, and their interconnections with other world basins through global trade and global economy.

THE SEINE BASIN

Today the Seine basin encompasses the major economic activities, except mining, that have increasingly put pressure on rivers and their basins over the last p40

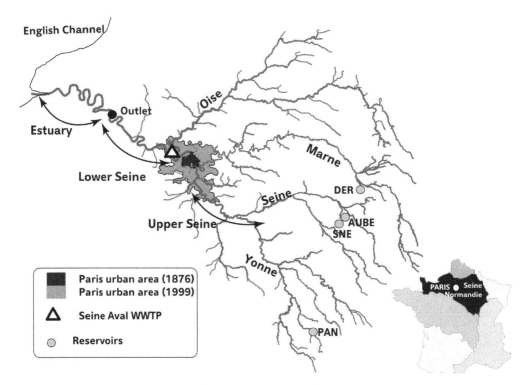

FIGURE 7.1. Main components of the Seine River basin and Paris urban growth between the 1870s and the 2000s. The river network corresponds to stream orders 3 to 7.

years: industrial, urban, agricultural, river transportation, and damming (Billen et al. 2007; Mouchel and Billen 2008–15).

The river basin upstream of its estuary covers 65,000 km². It had approximately 7 million people (Mp) in the 1870s; it has 17 Mp today. One important characteristic of the basin is the population pressure gradient, which has changed from fewer than 20 people/km² in half of the basin to more than 2,000 p/km² in the Paris suburban river basins (Orge, Bièvre), with an average of 250 p/km² for the whole basin at the mouth. Paris megacity is a prominent feature of the basin, which delineates the Upper Seine and the Lower Seine (fig. 7.1). The natural dilution power here is very limited, making the Seine sensitive to point sources of pollutions compared to most EU rivers. This is especially important given the fact that Paris's treated wastewaters total more than 30 m³/s, equivalent to a medium-sized river.

The Seine basin is also characterized by intensive agriculture. The development of agriculture at first paralleled the growth of Paris and its food demand over the last centuries (Billen et al. 2012). Then a major turning point in land use took place in the 1960s, when grasslands were converted to cropland and nitrogen fertilizers were used intensively. Further affecting the river has been the demand for deeper, larger, and more extended navigated reaches in the basin and the increased sand and gravel extraction in the floodplain, which has been used for Paris urban growth.

Industries are mostly located in Paris megacity, along the Lower Seine industrial corridors, and along one of its main tributaries, the Oise. Until the 1970s, industrial waste waters were barely treated on site and then discharged directly to the closest rivers, with the assumption that they would be diluted and self-cleaning. Until the late 1980s, the level of toxic substances in the river, the fluxes released by both the city and its industries and their effects on receiving waters, was largely ignored by French scientists and authorities (Meybeck et al. 2016).

Greater Paris (the Paris urban area) has evolved from 2.5 Mp over 480 km² in the 1870s to 10 Mp over 2,500 km² today (see fig. 7.1). In the 1870s, the collected waste waters started to be spread in sewage farms near Paris. Wastewater treatment plants (WWTP) were then gradually built in the past fifty years as a result of the 1964 Water Law. One of the sewage farms at Achères, located at 60 river km downstream of Paris, was converted between 1930 and the 1980s to the Seine-Aval WWTP, treating up to 8M equivalent-people in the 1970s. Since then, new WWTPs have been built around Paris (Lestel and Carré 2017).

PHYSICAL AND CHEMICAL IMPACTS ON THE RIVER NETWORK, A STREAM ORDER ANALYSIS, AND THE WEIGHT OF PARIS MEGACITY

The spatial representation of alterations to river and stream courses is difficult for several reasons: the increasing size of hydrological entities from headwater streams to the estuaries; the discrete nature of the information, collected at fixed stations; and the integration of water and sediment chemistry over the basin upstream of the station. In the classical way of representation, used by river authorities, stations are mapped as dots that are color-coded according to quality levels. This representation combines visually, and often statistically, the stations on small streams (basin area 100 km²) with those on great basins (100,000 km² or more). Stream orders, used by hydrologists and river ecologists (Naiman 1983), allow assessment of the quality of basins by their hydrological importance. The hydrological network is organized from the first permanent headwaters streams (order 1) to the river mouth (order 7 for the Seine). In many Piren Seine models the physical properties of the network, width, depth, water discharge, are considered similar within the same order and increase when two streams of similar n orders meet to form an $n+1$ order.

Figure 7.2A shows the distribution of the maximum physical alterations of the aquatic system by stream order. Unless otherwise noted, these date to 2015. The impacts are graded in four categories, according to their relative intensity on watercourses within a given stream order. These alterations have been gradually developed through time. In headwaters (orders 0 to 1 and 1 to 3) agriculture is the main cause of the alteration through wetland draining, ditch construction, and stream course straightening. The urban development of Greater Paris has generated the disappearance of many urban rivers, particularly at the turn of the nineteenth century (e.g., the Bièvre River). After 1950, channelization and dredging of the Seine River for navigation, excavation of dozens of large sand pits for gravel and sand mining, and regulation by locks are responsible for a major artificialization and regulation of the lower river reaches over several hundred kilometers, including the estuary. In addition, four main water reservoirs were constructed 250 river km upstream of Paris in the 1930s through 1980s (Pannecière [PAN], Seine [SNE], Aube [AUBE], and Der [DER]; see fig. 7.1). These were financed by the city of Paris for flood protection and low-water discharge regulation, increasing the summer low flows from 25 m^3/s at Paris up to 100 m^3/s for an increased dilution of treated Paris waste waters.

Other physical modifications also had an impact on orders 1 to 4 before the 1800s. These included multiple ponds—more than 2,550 for the whole Seine basin, mostly on first-order streams (69 percent) (Passy et al. 2012)—and water mills—up to 6,000 over 12,000 km^2 in the Ile-de-France region (Boët et al. 1999). The higher orders remained comparatively untouched and featured multiple islands. These islands, in turn, gradually disappeared between 1850 and 1950: in the 5 to 7 stream orders about 25 percent of the river bank length has been lost when comparing pre-1850 and contemporary maps (Lestel et al. 2015).

As such, the whole Seine River network is physically modified, with the exception of some forested streams. Meanwhile, land use has greatly evolved since 1950. For instance, in the middle reach of the Seine, upstream of Greater Paris, artificialized land cover (intensive agriculture, urban area, sand pits, channelized river course) increased from 51 to 74 percent, and more natural cover (forest, grassland) decreased from 49 to 26 percent. The sand pits excavated in the floodplain went from 0.1 to 7.6 percent (Bendjoudi et al. 2002). Mills, sills, and ponds can be considered semireversible features at secular time scales, but great reservoirs, loss of islands, channelization, and artificial embankments can be considered irreversible alterations that have modified the river ecology—for example, for fish (Boët et al. 1999; Tales et al. 2009).

The chemical (e.g., metals) and biogeochemical (eutrophication, hypoxia) alterations are here presented at their maximum stage (see references in Meybeck et al. 2016) (fig. 7.2B). Eutrophication developed when the river course was slowed down and/or in navigated reaches (stream orders 5 and up). Heavy metal (Cd, Cu, Pb, Zn) contamination is also organized by stream orders, the highest being the most degraded (Meybeck 1998, 2002). Small urban streams within Greater Paris did not follow the stream order progression as their high population was not always connected to treatment plants: they were more degraded than the Seine River itself. Also, in contrast to the general upstream-downstream degradation of the river chemical quality, following the population density distribution, the

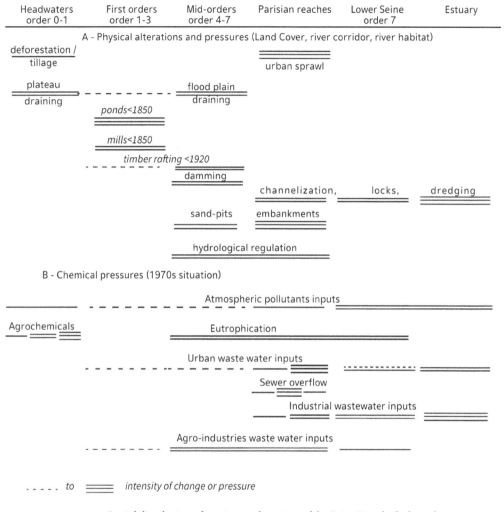

FIGURE 7.2. Spatial distribution of maximum alterations of the Seine River hydrological network, presented by stream orders 1 to 7. A: Physical alterations, as in 2015, otherwise noted. B: Chemical pressures as in the 1970s.

nitrate level in unpopulated streams draining intensive agriculture was already—and still is—at its highest level in the basin.

Our studies also reveal the historically enormous influence that Paris megacity has had on its river (Lestel and Carré 2017) (fig. 7.3). The hyperconcentration of population and industrial activities, and the subsequent releasing of their treated wastes from 1950 to the 1990s (Lestel and Carré 2017) (figs. 7.3, 7.1a), may have had impacts far downstream to the estuary. These include delayed nitrification (3) of

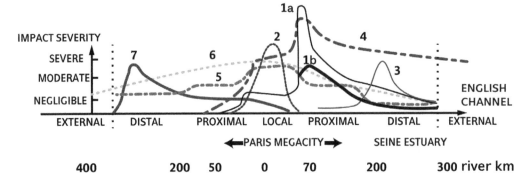

FIGURE 7.3. Schematic longitudinal profiles of the impacts of Paris megacity on the Seine River main course at various periods. 1a, 1b: Organic pollution in the 1960s and 1990s. 2: Occasional overflow of combined urban sewers (until the 1990s). 3: Estuarine hypoxia due to estuarine nitrification (until the 2000s). 4: Metal contamination (in 1990). 5: Physical habitat degradation (2015). 6: Atmospheric pollution (1980s). 7: Timber rafting (1600–1920)

the released ammonia with subsequent estuarine hypoxia (Garnier et al. 2007). This effect has been augmented by the fact that for several decades most collected waste waters were treated in a single location, the Seine-Aval WWTP (see fig. 7.1). In addition, many suburban wastes were discharged directly into the river, as was the case for industry. Today oxygen balance has been greatly improved (1b). Until the 1990s, during storm events, the impact of combined sewage overflow affected the Paris city center, generating hypoxia and fish kills (2). Authorities have since made a great effort to store these untreated waters then release them to WWTPs after the storm event. The metal level in particulates downstream of Paris was near its maximum in the 1970s (Meybeck et al. 2007) (4) and contributed to the general contamination of the English Channel and the North Sea.

Paris's impact is also observed in its distal upper course. Prevailing winds may carry atmospheric pollutants to other river basins (6). Water discharge regulated by its four major reservoirs actually constrains the river flows of the Yonne, Upper Seine, and Marne (5). For three hundred years until 1920, the Yonne-Seine River reach conveyed timber for fuel and construction wood to Paris (7); its impact on river ecology—wood debris, bank abrasion—has not yet been estimated.

ACCELERATED CIRCULATION AND OUTPUTS OF MATERIALS IN THE SEINE RIVER BASIN

River basins are traditionally used by geochemists and earth system scientists to establish the circulation of elements at the earth's surface in natural conditions and to understand its regulation. They determine (i) the natural composition of river solutes (mg/L, µg/L); (ii) the relative contents of elements in river particulates (% to parts per millions, or ppm; i.e., µg/g); and (iii) the exportation of these products by the river, rated by the basin area, also termed specific loads (mass per unit time and unit area: t km^{-2} y^{-1}). These metrics are used to quantify the natural earth system and reconstruct its past evolution in geologic eras.

The human impact on river fluxes has been recognized early, from the local to the global scale (Garrels et al. 1975; Meybeck and Helmer 1989; Berner and Berner 1996). Over the course of fifteen years, this field greatly expanded (Meybeck 2003; Vörosmarty and Meybeck 2004; Seitzinger et al. 2005; Vörösmarty et al. 2010; Syvitski and Kettner 2011), revealing major transformations of the earth's system on continents during the Anthropocene epoch: (i) the accelerated circulation of elements with regard to the preindustrial conditions, (ii) the retention of river particulates in the countless small to very large reservoirs built since 1950, and (iii) the related loss of water by irrigation, mostly in semiarid regions. The Seine River basin can be used to illustrate the river flux increase since retention in reservoirs is limited (Meybeck, de Marsily, and Fustec 1998). The Piren-Seine scientists have determined the evolution of river fluxes by combining several approaches (Meybeck et al. 2016): (i) the analysis of forested streams without any human impacts, for background levels of major ions and nutrients (Meybeck 1986); (ii) the analysis of Neolithic river floodplain deposits for background metal contents in river particulates (4000 BP, Meybeck et al., 2007); (iii) the reconstruction of the medieval circulation of nutrients in rural conditions (Billen et al. 2009); (iv) the reconstruction of river sediment composition, over the past eighty years, based on sedimentary archives in the Lower Seine floodplain (Meybeck et al. 2007; Le Cloarec et al. 2011) (see fig. 7.4, lower right cartoon); (v) the current circulation of nutrients and metals in the basin since 1950 through the compilation of economic data on fertilizer use, phosphorus use in detergents and other products, and the metal used in various sectors (as raw metal, metal containing products, recycled metals) (Meybeck et al. 2007; Thévenot et al. 2007; Lestel 2012; Billen et al. 2012; Garnier et al. 2015; Romero et al. 2016). In some cases, the data were only available at the national level, and a 30 to 40 percent proportion has been applied to convert those for the Seine basin—in proportion to its overall agricultural, demographic, and industrial weight. The limitations of these estimations are discussed by Lestel et al. (2007).

The river-borne fluxes at the river outlet (river budget station, monitored since the 1970s; see figs. 7.1 and 7.4) have been established for nitrogen, phosphorus, and heavy metals and compared to the economic flows of materials containing these elements over the 65,000 km^2 of the Seine basin. Several indicators are defined: (i) the *circulation ratio* of contemporary elemental circulation over natural (preindustrial) river flux: $I_1 = U_{eco}/F_{bgr}$; (ii) the *concentration ratio* of the contemporary concentrations or contents over the estimated natural levels $I_2 = C_{river}/C_{bgr}$;

TABLE 7.1. Indicators of the alteration of natural elemental fluxes in river basins, resulting from a mass flow and river flux comparison. The Seine River example. I_1 to I_4: see text. WQC_1/C_{BGR} water quality criteria over background concentration, established by geochemists.
(1) River fluxes based on dissolved material around 2000s. (2) River fluxes based on particulate matter at the maximum contamination period (ca. 1960). (3) In g capita^{-1} y^{-1}. DL: dimensionless ratio. WQC_1/C_{BGR} defines the deviation from the pristine state accepted by river managers (Ministère de l'écologie 2012; Oudin and Maupas 1999 for metals) (see fig. 7.5).

	Nitrate NO_3^-	Phosphorus P	Cadmium Cd	Copper Cu	Mercury Hg	Lead Pb	Zinc Zn
I_1 (DL) (max)	40 (1)	280 (1)	3000 (2)	13000 (2)	5000 (2)	6000 (2)	2000 (2)
I_2 (DL) (max)	(25) (1)	(50) (1)	150 (2)		500 (2)	23 (2)	22 (2)
I_3 1960s (3)			3.3	39.6	0.61	52	156
I_3 2000s (3)	4400	400	0.07	5.3	0.06	5	11.1
I_4 (%) 1960s			11.6	0.35		0.94	1.9
I_4 (%) 2000s	15	7	0.4	0.04	(10)	0.08	0.25
WQC_1/C_{BGR}	10 to 50	3 to 5	4	1.8	23	1.1	2.3

(iii) the *per capita excess loads* in the river $I_3 = (F_{riv} - F_{bgr})/Pop$, calculated by subtracting the natural exports at river mouth (F_{bgr}) from the measured or reconstructed river loads (F_{riv}) at given periods, defining excess loads, then rating it to the basin population (Pop) during these periods (expressed in g capita^{-1} y^{-1}); (iv) the *leakage rate*, that is, the ratio of excess river load (annual mass) to the elemental circulation (annual mass) over the basin, $I_4 = (F_{riv} - F_{bgr})/U_{eco}$ (Table1); (v) the ratio of the contemporary water quality criteria defining the good state over the natural background (WQC_1/C_{BGR}).

As none of these indicators is affected by the size of the basin, they allow making comparisons between river basins and elements, particularly as concerns I_1, I_2, I_4, which are dimensionless. The (I_1) indicator, expressing the flow of economic materials with regard to natural processes in the earth system within a river basin territory, ranges here from 40 to 13,000. The concentration ratios (I_2) measure the rate of deviation of concentrations from the pristine river state, an indicator often used by geochemists, which reached maximum values from 20 (nitrogen, zinc) to 500 (mercury). It depends on the natural dilution power of the receiving river: for a given pressure, for example, a great city, I_2 is lower when the receiving river has a higher water discharge or sediment flux, as with the Rio Negro for Manaus and the Rhône River for Lyon, respectively; in the Yellow River it is barely possible to find evidence of metal contamination, due to the enormous sediment load of the river, a thousand times that of the Seine. The *per capita* excess loads (I_3) depend on the use of material, on the efficiency of the environmental responses (e.g., recycling and water treatment). Between the 1960s and the 2000s, they have been divided tenfold for Cu, Hg, Pb, and Zn, and by fifty-fold for Cd, the use of which is now greatly restricted. The per capita river export of nitrate-nitrogen is eleven-fold that of phosphate-phosphorus. The leakage rate to the aquatic system (I_4) is an indicator of the environmental performance of the society within the basin (Meybeck et al. 2007). This is still important for N and P but is today very limited for most metals (from 0.04 to 0.4 percent), except mercury, which still affects the river despite its complete ban for most uses. Within the WFD, the management of river basins, targeted on concentrations that define the "good ecological state" (WQC_1), may not reflect the environmental efforts, better described by I_3 and I_4. It must be noted that the WQC_1 set by French water authorities are often much higher than the natural background values (C_{BGR}) as estimated by scientists (see table 7.1 below), and that current water management is not based on the environmental impact indicator, I_2, or on the indicators of environmental performance, I_1, I_3 and I_4.

THE OPENING OF THE SEINE BASIN DEMONSTRATED BY THE MATERIAL FLOW ANALYSIS

For earth system scientists, natural fluxes within river basins are derived from the erosion and weathering products of surficial rocks and from the uptake of atmospheric carbon and nitrogen occurring within the basin. For ecological economists, the material flow analysis over a given territory reveals the metabolism of the anthroposphere (Baccini and Brunner 2012). The comparison of both approaches reveals that the circulation of many economic products used in the Seine River basin is one to two orders of magnitude more than the natural fluxes, as for the heavy metals (Lestel 2012). Most of these products are actually recycled, and the river is receiving a minor leak of these. Also, the economic circulation of products in the Seine River basin is now totally opened: the basin exports a great quantity of agricultural and food products and manufactured products and consumes a large quantity of fossil fuels, mining products, and manufactured products; it also emits long-range atmospheric pollutants that may reach other basins.

The material flow analysis of heavy metals in the Seine basin is schematized in figure 7.4. It illustrates the spatial and temporal complexity (Lestel et al. 2007; Thévenot et al. 2007; Lestel 2012): mining (1) only occurred in the 1700s, and inherited contamination is expected in floodplain sediments of the Upper Seine and Marne (see fig. 7.1); metal smelting (2) was located throughout most of the nineteenth and twentieth centuries in Paris and along the rivers (Oise, Seine) (Lestel 2012); metal transformation by industries (3) is now located mostly in Greater Paris; the use of metal and of metal-containing products is very much related to the urban population (4); the recycling of metal products, such as pipes and car batteries, a great provider of metal leaks, was first realized in Paris suburbs prior to 1950 (5a) and then externalized outside of the Seine basin (particularly in the north of France, where it generated extreme contamination, and finally outside the country) (5b). The state of contamination of river reaches generated by these activities depends on their location and on the ratio pressure/river dilution power.

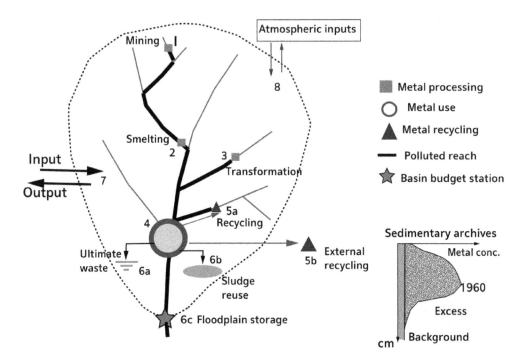

FIGURE 7.4. Schematic representation of the circulation of material within a river basin at the Anthropocene and reconstruction of past contamination from floodplain sediments at the basin outlet. The metal example. See text for explanations.

Since most of the heavy metal flux is associated with particulates, their storage over centuries can be important in urban infrastructures (4), in soils contaminated by industries, in former landfills and dredged sediments (6a), in agricultural soils fertilized by the recycling of Paris WWTP treated sludge, once highly contaminated (<1990) (6b), and in floodplain sediments, which provide records of the contamination (6c).

Changing biogeochemical cycles are also illustrated by nitrogen circulation. For centuries, Paris city growth depended on its fertile hinterland. Organic wastes from animal and human populations were collected and recycled in suburban market gardening, at a short distance from the city. The use of industrial fertilizers, mostly after the 1950s, and the development of sewage collection and treatment reduced the recycling loop of carbon, nitrogen, and phosphorus (Barles 2005; Billen et al. 2009; Billen et al. 2012), and is recently being reconsidered.

Finally, global trade should now be considered in the analysis of environmental impacts by societies (figs. 7.4, 7.7). For instance, today imported products are present in all sectors: soya food for cattle, tropical fruits and grains, palm oil, tropical woods, fuels, metallic ores, metal-containing products, and so on. These generate substantial environmental degradation where they originate but are rarely taken into account in the environmental assessment of the receiving river basins. In turn, the nitrogen circulation at the global scale shows that the export of food produced within the Seine basin is a significant nitrogen input into other basins (Lassaletta et al. 2014). Atmospheric exchanges (8) may also be considered in river basin budgets.

REVERSIBILITY AND IRREVERSIBILITY OF RIVER QUALITY IN THE LOWER SEINE RIVER BASIN

Analysis of the physical, chemical, and biochemical impacts that occur in the Seine River over the longue durée shows various trajectories (Meybeck et al. 2016), some reversible, others irreversible, that can be schematized using the "impair-then-repair" model (Meybeck, de Marsily, and Fustec 1998; Vörösmarty et al. 2015) (fig. 7.5). This model starts with a period of *insignificant impact* on the earth system or on the water resources (OA, stage 1). The next stage is a period of *accelerated degradation* of the aquatic environment and of water resources (AB, stage 2), often faster than population increase in the river basin, reaching first the level of water quality (WQC_1) at which water resources are impaired, then often followed by a severe level of water quality (WQC_2). When the technical and regulatory measures taken by a society become efficient a *proactive rehabilitation phase* is observed after a maximum impact stage (BC, stage 3). When a satisfactory state is finally achieved, reaching the level of quality WQC_1, the *regulation stage* ensures a stable quality even if the population and economy of the basin continue to grow (CD, stage 4). In the Seine River, the impact of untreated wastewater combined with sewer overflow—a historical heritage—is now minimized by the management of sewage works during storm events in Greater Paris (Tabuchi et al. 2013). If environmental management is insufficient, the impact can reach *permanent degradation* (BE, stage 5) stabilized at an altered level (>WQC_2), and the change can be considered irreversible.

The duration of the moderate environmental degradation (ED_1), from the societal perspective, is defined here by the exceedance of WQC_1, ; and the duration of the severe degradation (ED_2), by the exceedance of WQC_2. Water quality scales, WQC_1 and WQC_2, defined for each of the issues recognized by the society may evolve over time, therefore changing the environmental assessments it makes. From an earth system perspective, the analysis may be quite different: any significant deviation from the pristine state, as defined by the background concentrations (C_{BGR}), is expressing an alteration of the earth system (ES_A) and may generate a change in receiving waters—for instance, along the coastal zone. The WQC_1/C_{BGR} figures (table 7.1) in which WQC_1 is the contemporary threshold of the good ecological state used by French authorities, range from 1 to 50, depending on elements: they are much higher for nitrate than for metals, reflecting differing societal perspectives on the most toxic substances.

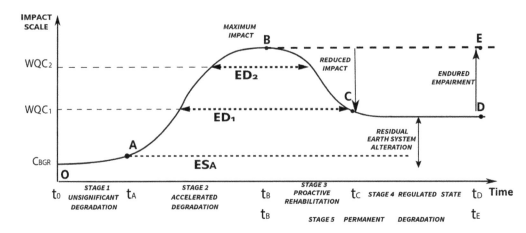

FIGURE 7.5. The impair-then-repair scheme and the five stages defining river quality trajectories, applied to North American and Western European river basins (adapted from Meybeck 1998; Vörösmarty et al. 2015). WQC1 and WQC2: Water quality criteria established for water management. CBGR: Pristine state concentrations. ESA, ED1, ED2: Respectively, duration of Earth System alteration, of the impaired state and severe degradations of the river, as defined by river basin societies.

In the Seine River basin this scheme has been applied to analyze the 1870–2010 trajectories of some river quality issues, including physical alterations, nitrate pollution, eutrophication, "organic pollution" (leading to hypoxic waters), heavy metals contamination, and bacterial contamination. Many of them are detailed in a companion paper (Meybeck et al. 2016).

The physical alteration of the Seine River network, at any stream order, is mostly irreversible (>100 years), and therefore at stage 5 (see fig. 7.2). However, ancient mills and ponds are semireversible: many fish ponds have been filled from 1800 to 1950, and the many sills left by water mills have been dismantled in response to the Water Framework Directive, which favors fish circulation (stage 4). In contrast, the major water works (locks, channelization, artificial banks, sand pits, dredged reaches and reservoirs) have barely been studied and can be considered irreversible changes that have generated the loss of five migratory fish species in the basin. This alteration developed gradually between the late Middle Ages (t_A, fig. 7.5) and 1990 (t_B), when the last reservoir was constructed. The most critical period (ED_2) started in the 1900s when the salmon disappeared.

The chemical alterations of solutes—dissolved nutrients, organic matter—are reversible, provided that adapted technical or regulatory responses are applied (Meybeck et al. 2016). For organic pollution (river hypoxia, ammonia pollution), the river is now reaching stage 4. The maximum hypoxia period is observed at least between the 1870s and 1990s (ED_2), until the completion of WWTPs in the whole basin. As concerns river eutrophication, controlled by dissolved phosphate, the river network is currently at stage 3–4 and the ED_2 period extends from the 1960s to the 2000s, when detergent phosphates were finally banned and tertiary treatment of phosphorus was established in Seine-Aval WWTPs and others. The nitrate pollution issue is still highly debated by basin actors, depending on the perspective considered and the WQC_2 chosen by the society. The impaired state (ED_1, 1960–present) started after deep changes in land uses and agricultural practices. Since then, the nitrate concentration has gradually increased and reached a river maximum near 2000, after which it has remained stable. From the sanitary perspective, which prevailed for the basin authorities from the 1970s to the 2000s, the current situation in the river is not severe: nitrate is around 25 mg/L, compared to the 50mg/L WHO guideline; however, many wells have exceeded this guideline and have been closed. From the point of view of coastal eutrophication, bathing and seafood consumption, nitrates are much too high, producing green tides and toxic algal blooms, and the river threshold established by coastal scientists is 10mg/L. From this perspective, the river has been at stage 5 since 1970, and stage 3 will not be initiated unless drastic changes in fertilizer use are made. And even with such changes, the nitrate contamination of groundwater may last half a century (Meybeck et al. 2016). The 10mg/L threshold still corresponds to a tenfold increase in nitrate load as regards pristine conditions, an increase factor that cannot be accepted from an earth system perspective for oceans such as the North Atlantic.

The bacterial contamination trajectory in the Middle Seine River, upstream of Paris city, can be assessed, thanks to very early surveillance in the 1900s. It shows a marked degradation between the 1950s, when sewage collection was generalized in Paris suburbs, and the 1990s, when treatment capacity for Paris and its suburbs became sufficient (Servais et al. 2007) (stage 4). For the Lower Seine the ED_2 period extended for more than 140 years, and the trajectory is currently at stage 3–4.

The chemical alteration of heavy metals, as measured in river particulates, shows irreversible impacts in contaminated soils and floodplain sediments. Their levels greatly depend on sediment quality criteria that have been divided up to tenfold since the 1980s. The general trajectory of metal contamination as archived in river sediments (see fig. 7.4) is a general decrease in content since the 1960s (t_B, fig. 7.5) in the Seine basin, mostly due to industrial transformations (Meybeck et al. 2007; Lestel 2012). The duration of severe degradation (ED_2) was at least from the 1920s to the 1990s for mercury and cadmium, though this duration was shorter for copper, lead, and zinc. Today metal levels in river particulates have decreased (stage 3 or 4, depending on metals) but are often much higher than the pristine state established by using 4,500-year-old sediments. This is particularly the case for mercury. The Seine River has been and continues to be a major source of metal contamination of the North Sea. Contaminated sites in soils and sediments will last for millennia, unless specifically addressed.

Other micropollution issues are now addressed in the Piren-Seine program (endocrine disruptors, drugs, pesticides): most of their trajectories correspond to stage 2. The above description of the historical impacts of Paris on the Seine River has parallels in other cities, including Berlin, Milan, and Brussels (Lestel and Carré 2017). The severe degradation period (ED_2) of the Spree, Lambro, and Zenne Rivers also lasted for one hundred years for most issues. We should expect similar impacts in many other fast-developing megacities.

CONCLUSION

Although the "good state" targeted in Europe (WFD 2000) leads us to believe that the goal of society is now to bring the river basin back to its natural conditions, this is by no means the case from an earth system perspective (fig. 7.6). First, many changes are already (or are becoming) irreversible—at least for many generations with regard to land use changes (deforestation and agriculture, urban growth, reservoir construction); degradation of the aquatic habitat throughout the river continuum, from headwaters to estuary; soil and sediment contamination; and aquifer pollution. Second, the circulation of elements in such river basins can be greatly modified.

From the earth system perspective, the natural circulation of elements within river basins might be multiplied by more than one order of magnitude when megacities are present. Leaks into river basins can range from 0.1 to 10 percent, depending on the elements and the societies' stages of development, and can dramatically modify natural concentrations and fluxes in rivers. Leakage rates are not stable: metals in the Seine River basin have decreased over the past fifty years, by one order of magnitude. This was due first to important changes in the industrial sector and then to environmental regulations—even while the use of most metals was increasing (Lestel 2012; Meybeck et al. 2007). Meanwhile the per capita excess loads carried by the river have decreased, reflecting both economic development and environmental responses.

From the perspective of water resources used by societies, human impacts on river basins should be analyzed from multiple points of views, considering the spatial heterogeneity and the multiple trajectories over the longue durée. For instance, the stream order approach should be complemented by the upstream/downstream impacts of the megacity. Past activities (mining, industrial, urban, and even agricultural cadmium-containing phosphorus fertilizers) have a bearing on present contaminations of river and soil particulates. Cumulative past alterations of the physical habitat, started one hundred years ago or more, undertaken to secure water resources, meet navigation needs, develop agriculture, and flood security, are mostly permanent and generate irreversible impacts, such as the loss of migratory fish communities. In contrast, organic pollution, river eutrophication, and bacterial contamination have the potential to be repaired.

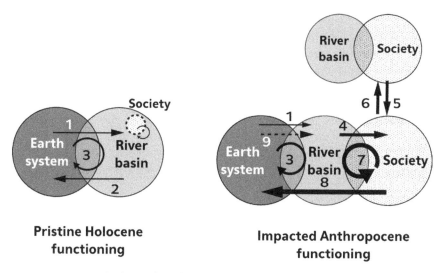

FIGURE 7.6. General scheme of circulation of material within pristine basins (left) and impacted basins at the Anthropocene. See text. Atmospheric transfers are not presented here.

Further, the impact of a megacity on a river basin is not necessarily proportional to its population: at some stage of environmental concern, economic and technical development, societies may improve the recycling rate of their economic materials, waste collection and treatment, modify the use of toxic substances (arsenic, mercury, cadmium, atrazine herbicide in the Seine basin), and improve markedly the chemical and biogeochemical quality of their rivers. The river biota, also exposed to species introductions and invasions, reflect these multiple impacts and their trajectories. Thus river-society interactions are spatially and temporally complex and can be addressed only by means of a multidisciplinary approach in which contemporary hydrologists, geochemists, hydrobiologists, and geographers are collaborating with environmental historians (Lestel and Carré 2017).

The evolution from natural to Anthropocene conditions is hypothesized in figure 7.6. The nonindustrial agrarian society is biogeochemically in equilibrium with the basin resources mostly controlled by the earth system (fig. 7.6, left, 1). The fluxes exported by rivers to oceans (2) contribute to the general balance (3) of the earth system. The Anthropocene is characterized by the globalization of river fluxes and their controls (fig. 7.6, right). In populated and industrialized river basins, the circulation of economic materials (7), extracted within the basin (4) and/or imported (5), may exceed by an order of magnitude or more the natural state (1) (e.g., nitrogen, phosphorus, and heavy metals in the Seine basin), resulting in additional river fluxes and concentrations (8). The most developed basins affect the least developed ones, sometimes far away, to meet their own mining, agricultural, or energy needs (e.g., hydropower dams) and may transfer their own wastes to these basins.

Environmental measures aim to minimize the leaks from the anthroposystem to the river system (8) but actually accept a variable level of deviation from the pristine state (WQC$_1$ vs. C$_{BGR,}$ table 1; stage 4, fig. 7.5); the long-term impact (centuries to millennia) is still poorly known. The regional and global water and sediment fluxes to oceans are already modified, and the most sensitive biogeochemical cycles—nitrogen, phosphorus, silica—are likely to modify the earth system and, in turn, generate global changes that will affect all river basins (Meybeck 2003; Seitzinger et al. 2005; Vörösmarty et al. 2010; Syvitski and Kettner 2011).

ACKNOWLEDGMENTS

This work has been generated by the Piren-Seine program, initiated in 1989 by Ghislain de Marsily, then directed by Gilles Billen and Jean Marie Mouchel, and now by Nicolas Flipo. We have used the work of dozens of our colleagues and benefited from our annual meeting discussions. The position of river basins within the Anthropocene started to be analyzed in the IGBP with colleagues in the BAHC, LOICZ, and IGBP-Water group, particularly Charles Vörösmarty. The water quality issues are addressed in the French ANR Makara program (ANR-12-SENV-009). Aurélien Baro assisted us in preparing the illustrations, and Jason M. Kelly provided editorial advice, although we might have added other editorial flaws. To all we express our warm thanks.

8

Anthropocene World / Anthropocene Waters

A Historical Examination of Ideas and Agency

Philip V. Scarpino

When I was in high school in the first half of the 1960s, I was fascinated by science fiction. The concept of terraforming was one of the key themes in the science fiction that I read. In the fantastic and fanciful worlds created by science fiction writers, human beings employed science and technology and energy to refashion (or terraform) the hostile environments of alien planets to support human life. The word *terraform* first appeared in print in July 1942. A writer named Jack Williamson employed it in an article titled "Collision Orbit," published in a magazine called *Astounding Science Fiction.* In the early 1950s, the great trio of science fiction writers, Robert Heinlein, Arthur C. Clarke, and Isaac Asimov, adopted and used *terraform* in a way that influenced popular culture (Heinlein 1950; Clarke 1951; Asimov 1952; Fogg 1995). By the early twenty-first century, a descriptive term coined by a science fiction writer and published in a science fiction pulp magazine in 1942 would be superseded by a concept generated by one of the world's leading atmospheric scientists—a concept that would highlight the dominant role played by human beings in fundamentally transforming (or terraforming) the environment of our own planet Earth.

Use of the word *terraform* by leading science fiction writers in the 1950s corresponded with a widespread faith in science and technology and cheap, abundant fossil fuel and natural resources to solve pressing social problems and improve the quality of life here on earth. Large, American-made automobiles powered by gas-guzzling internal combustion engines represented a material symbol of the good life that resources and energy and industrial production could provide.[1] Seemingly amazing products of organic chemistry offered technical fixes for pressing issues that had long plagued human populations. Petroleum-based synthetic organic pesticides promised to rid humanity of the scourge of insects that spread disease and ruined crops; combined with synthetic organic fertilizers and herbicides these new compounds held out the potential for a green revolution that would allow fewer farmers to feed ever more people. Chlorofluorocarbons used as refrigerants made in-home refrigerators much safer and facilitated the routine use of air-conditioning, which in turn enhanced comfort levels for tens of millions worldwide.[2] More food and better public health contributed to a rising global population.

Fast forward to the early 1970s—on December 7, 1972, just past the peak of the popular, ecology-based environmental movement, *Apollo 17* sent back the classic "Blue Marble" photo, showing the earth wrapped in its envelope of atmosphere hanging in the black vastness of space.[3] As we tentatively moved out into space, one of the most inspiring outcomes was to look back and gain a new angle of vision on our own earth. At approximately the same time, the research of atmospheric scientists undertaken in the late 1960s and the first half of the 1970s eventually contributed to a significant, new way of seeing our "Blue Marble." Life-sustaining atmospheric systems proved vulnerable to consequences of human action, including but not limited to burning fossil fuel in internal combustion engines and the widespread use of artificial fertilizer and chlorofluorocarbons.

Paul Crutzen earned his PhD with highest distinction in 1968 at the Meteorology Institute, Stockholm University, writing a dissertation titled, "Determination of Parameters Appearing in the 'Dry' and the 'Wet' Photochemical Theories for Ozone in the Stratosphere." In 1970, Crutzen published an important article in which he referenced earlier research reporting that nitrous oxide (N_2O) likely produced naturally by bacteria in the soil could influence the levels of nitrogen oxides (NO and NO_2) in the stratosphere. Building on those findings, Crutzen observed that "the NO and NO_2 concentrations have a direct controlling effect on the ozone distributions in a large part of the stratosphere, and consequently on the atmospheric ozone production rates" (Crutzen 1970, 320). Crutzen's findings were poised to become one of two important streams of research that established links between human agency and a relatively small but crucial layer of ozone high in the stratosphere that protected people and most other life on earth from the potentially harmful impact of the sun's ultraviolet rays. Ultimately, that research would not only transform scientific understanding of atmospheric systems, but also held the potential to revolutionize the ways in which human beings understand their relationship with the earth's environment.

In subsequent publications, Crutzen postulated that anthropogenic emissions from increasing use of artificial fertilizer and high-flying supersonic aircraft might add to the levels of nitrogen oxides in the stratosphere, and augmented levels of nitrogen oxides could deplete the earth's crucial ozone layer. "It has been indicated during recent years," Crutzen argued in 1974, "that important reductions in atmospheric ozone may be caused by a number of human activities such as stratospheric aviation, increased use of nitrates as fertilizers and the use of chlorofluoromethanes (mostly known under the name 'freons')" (Crutzen 1974, 201; see also Crutzen and Ehhalt 1977). Crutzen's research highlighted connections between single-purpose technologies that may have fulfilled their primary purposes very well and unintended or unanticipated consequences that produced adverse impacts on the stratospheric ozone layer. Reflecting on his choice of a research topic, Crutzen explained, "I wanted to do pure science related to natural processes and therefore I picked stratospheric ozone as my subject, without the slightest anticipation of what lay ahead" ("Paul J. Crutzen—Biographical" 1995).

In the early 1970s, another, related stream of atmospheric research emerged that called attention to the harmful effects of a common and widely used refrigerant on stratospheric ozone. Chlorofluorocarbons (CFCs), which are synthetic organic compounds composed of carbon, fluorine, and chlorine, were first synthesized in the United States in the late 1920s as a safe alternative to chemicals then widely used as coolants in refrigerators. After World War II, CFCs (usually sold under the trade name Freon) came into widespread, worldwide use as propellants in aerosol containers, coolants in air conditioners and refrigerators, and solvents (Elkins 1999).

In 1974, Crutzen read a draft research report on the potential adverse impact of chlorofluoromethanes (marketed as Freon 31) on the ozone layer coauthored by Frank S. Rowland, a chemistry professor at the University of California, Irvine, and Mario J. Molina, a Mexican national working with Rowland as his postdoctoral associate (European Space Agency n.d.). The research by Roland and Molina revealed that Freon, which was stable and inert in the lower atmosphere, broke down in the stratosphere and released chlorine, which destroyed ozone. Crutzen responded to their research by examining a closely related compound and developing a model of the ozone depletion that could result from continued use of chlorofluorocarbons. His research yielded a sobering prediction: "up to 40% of ozone would be depleted at the 1974 rate" (European Space Agency n.d.).

Research published in 1985 by three scientists working for British Arctic Survey Stations revealed seasonal drops in stratospheric ozone above Antarctica likely caused by the action of chlorine associated with CFCs. The scientists themselves demonstrated an abundance of professional caution as their investigations moved forward, one of them arguing in 1987 that "the evidence implicating total chlorine, and hence the CFCs, remains circumstantial. It should, nevertheless, be heeded until more direct evidence can be obtained" (Farman 1987, 644; see also Farman, Gardiner, and Shanklin 1985). Their discovery of what quickly became known as the "ozone hole" added to accumulating evidence of a negative connection between widespread use of CFCs and ozone depletion.

The findings of Crutzen, Rowland, and Molina, as well as scientists associated with the British Arctic Survey Stations, contributed directly to the Protocol on Substances that Deplete the Ozone Layer, signed in Montreal, Canada, in September 1987 and "entered into force" on January 1, 1989. When combined with several subsequent adjustments between March 1991 and May 2008, the Montreal Protocol led to strict worldwide controls on CFCs and other ozone-depleting compounds (UNEP Ozone Secretariat 2016; Elkins 1999). In 1995, Crutzen, Rowland, and Molina shared the Nobel Prize in chemistry for their findings on ozone depletion. Their scholarship helped focus scientific attention on the powerful and significant impact of human activities on earth systems. It also turned out to be one of the few successful worldwide responses to the environmental consequences of human actions to take place in the late twentieth century.

At least in the case of CFCs and the "ozone hole," cutting-edge scientific research also caught the attention of a broad international public and officials of the forty-six nations that signed the Montreal Protocol (UNEP Ozone Secretariat 2016). Looking ahead to other international and worldwide environmental problems, such as access to and distribution of freshwater and climate change, "fixing" the ozone hole proved to be deceptively simple. Complex and complicated atmospheric science could be boiled down to a relatively easy-to-understand, near-term, and direct cause-and-effect problem: CFCs and related compounds were destroying the essential ozone layer, which in turn would have a significant, measurable, and detrimental impact on the health of human beings worldwide. And the solution was a relatively "simple" technical fix that did not require people to effect any significant changes in values and expectations or to accept alterations in lifestyle or standard of living. Political leaders who lined up behind the Montreal Protocol and elimination of CFCs did so knowing that they faced a very low risk of backlash from their constituents. All that was required was to substitute a new chemical refrigerant for Freon without any corresponding need to cut back on air-conditioning or anything else.

Fast forward again, to the early twenty-first century. Between 1800 and 2011, the earth's population increased from 0.98 billion to 6.9 billion, with the most rapid increases taking place in the past century. In 1950, when Robert Heinlein and Arthur C. Clarke and Isaac Asimov wrote about terraforming distant planets, the world's population stood at 2.52 billion. In 2013, the world supported more than 7 billion inhabitants. A pronounced trend toward urbanization has accompanied explosive population growth. In 1950, 29.4 percent of the world's population resided in cities. By 2011, the percentage of the world's population living in cities had risen to 52.1—with a clear developmental trend being concentration in ever larger cities (U.N., Department of Economic and Social Affairs, Population Division 2012, 4–6). Rapid population growth and urbanization pose serious challenges for access to safe, clean freshwater and for disposal of waste- and storm water runoff.

Population growth and urbanization in the past century were facilitated by a dramatic shift from solar energy to fossil fuel and a massive increase in the use of energy. Climate change stands at the head of the list of the unintended and unanticipated consequences of burning all of that fossil fuel in the atmosphere—illustrated at least in part by a rapid rise in atmospheric carbon dioxide (Intergovernmental Panel on Climate Change 2013b, 2). Growth patterns of population, energy use, and carbon dioxide reveal two important and interrelated historical trends: (1) the pace of change accelerated rapidly in the past one hundred years; and (2) most of the key variables that illustrate and reflect changes in earth systems follow an exponential growth pattern. It is worth noting that the long, slow period of "approach" to the "elbow" of an exponential curve represents an important part of the historical/developmental trend of the variable in question.

In 2000, Crutzen coined the term "Anthropocene" to describe a new geological epoch in which human action had become the primary driver of environmental change. According to Fred Pearce writing in *With Speed and Violence: Why Scientists Fear Tipping Points in Climate Change* (2008), Crutzen told him:

> I was at a conference where someone said something about the Holocene, the long period of relatively stable climate since the end of the last ice age. . . . I suddenly thought that this was wrong. The world has changed too much. So I said: "No, we are in the Anthropocene." I just made up the word on the spur of the moment. Everyone was shocked. But it seems to have stuck. (Pearce 2008, 44)

Crutzen's towering scientific reputation bolstered by his Nobel Prize instantly conferred a high level of authority and credibility on his declaration of the Anthropocene. It is not at all surprising that the term and its initial use originated with scientists who addressed their research to human impacts on global atmospheric systems, including climate change. After all, the Anthropocene refers to new sets of circumstances where the results of human actions impact global environmental conditions and actually produce a stratigraphic record. The term "Anthropocene" rapidly and informally entered the scientific literature, used to emphasize the dominant role of human activity in shaping the global environment (Zalasiewicz et al. 2008 ; Andersson, Mackenzie, and Lerman 2005; Crossland et al. 2006; Steffen et al. 2004; Syvitski et al. 2005). Through the lens of the Anthropocene, the boundaries between natural and human history blur; understanding the present-day environment requires paying as much attention to human agency over time as it does to the evolutionary trajectory of natural processes.

Species extinction represents a global phenomenon that has left distinct fossil evidence that can be identified in the stratigraphic record. In the past 540 million years, the earth has experienced five periods of mass extinction when at least 75 percent of the estimated species comprising earth's biota disappeared. While it is likely that each of the "Big Five" extinctions was precipitated by different causes, they all had at least two things in common. First, we know about these episodes of mass extinction by studying fossil evidence originally deposited in layers of sedimentary rock. The fossil record in effect serves as the "database" or the "archive" that documents the evolution of life on earth. Second, mass extinctions one through five took place in the complete absence of human agency.

Within the past few decades, scientists have begun arguing that earth may be entering a sixth period of mass extinction—driven directly by the actions of people.[4] Some of the new information, and especially that aimed at public audiences, declares that this sixth mass extinction is already under way. In the fall of 2014, National Public Television in the United States broadcast a documentary film titled *From Billions to None: The Passenger Pigeons' Flight to Extinction* (Mrazek 2014), which follows the naturalist Joel Greenberg, author of *A Feathered River Across the Sky: The Passenger Pigeon's Flight to Extinction* (2014). At the time of European contact passenger pigeons in North America may have numbered 3 billion to 5 billion. On September 1, 1914, the last known passenger pigeon died alone in the Cincinnati Zoological Garden. At a pivotal point in *From Billions to None,* David E. Blockstein, senior scientist at the National Council for Science and the Environment, makes the following observation about extinction:

> The driving force is now humanity; changing the forces of nature. And, one of the consequences of the way that we are driving everything on the planet is that we are driving so many of the other species—our fellow inhabitants of spaceship earth—we are driving them to extinction. And, the rate is unprecedented. There have been mass extinctions in historical times, but essentially we are like the asteroid that killed the dinosaurs and the impact that we have is as swift and as overarching as that asteroid that killed the dinosaurs. (Mrazek 2014)

The asteroid that killed the dinosaurs offers a compelling metaphor for human influence on earth systems, while the reference to "spaceship earth" calls up images of the "Blue Marble," now profoundly and directly threatened by the actions of its own human inhabitants. At the same time, Blockstein's comparison of humanity to an extinction-producing asteroid lacks the precision and evidence-based caution that usually characterizes professional, scientific publication.

A measured and professionally cautious article titled "Has the Earth's Sixth Mass Extinction Already Arrived?," published in *Nature* by Anthony D. Barnosky et al. in March 2011, takes on the question posed in the title of the article. Barnosky and his coauthors begin by noting that of the approximately 4 billion species that have evolved on earth in the past 3.5 billion years, about 99 percent have gone extinct. In the history of life on earth extinction is common, but under ordinary circumstances "speciation" balances loss. The article mentions the five periods of mass extinction evidenced in the fossil record and then turns to the question of a sixth episode caused by human action. Barnosky et al. explain the possibility of such a sixth mass extinction in the following anthropogenic terms:

> Increasingly, scientists are recognizing modern extinctions of species and populations. Documented numbers are likely to be serious under-estimates, because most species have not yet been formally described. Such observations suggest that humans are now causing the sixth mass extinction, through co-opting resources, fragmenting

habitats, introducing non-native species, spreading pathogens, killing species directly, and changing global climate. (Barnosky et al. 2011, 51)

The authors go on to explain that mass extinction, "in the conservative paleontological sense, is when extinction rates accelerate relative to origination rates such
that over 75% of species disappear within a geologically short interval—typically
less than 2 million years, in some cases much less." They conclude that recent historical extinction rates are both dramatic and serious, but they do not yet rise to
the paleontological definition of mass extinction. They also warn that loss of species in the "critically endangered" category "would propel the world to a state of
mass extinction that has previously been seen only five times in about 540 million
years." Further loss of species categorized as "endangered" and "vulnerable" could
bring on a sixth mass extinction in a few centuries. Understanding the difference
between the present extinction-related situation and where we could be in a few
generations "reveals the urgency of relieving the pressures that are pushing today's
species towards extinction" (Barnosky et al. 2011, 56; see also De Vos et al. 2015;
World Wildlife Fund 2014, esp. chap. 1; Monastersky 2014). Thus Barnosky and
colleagues argue that while the world has not yet entered a sixth period of mass
extinction, we are traveling toward a tipping point—only this time human actions
can either push life on earth over the edge or effect a change of course to avert the
looming disaster.

Construction of dams across rivers and streams offers an additional example of
environmental change that holds the potential to alter the sedimentary and eventually the stratigraphic record. According to a recent article by Katherine J. Skalak
et al. titled "Large Dams and Alluvial Rivers in the Anthropocene," "one of the
greatest modifications of the fluvial landscape in the Anthropocene is the construction of dams." Worldwide, the inventory of dams stands at about 800,000. All
of these dams have "increased the mean residence time of river waters from 16 to
47 days and has increased the volume of standing water more than 700 percent."
Construction of dams worldwide accelerated markedly starting in the 1950s and
peaked in 1968. "Large Dams and Alluvial Rivers in the Anthropocene" focuses on
the Garrison and Oahe dams on the main stem of the Missouri River in North and
South Dakota, examining the interactive and combined effect of dams constructed
in sequence on the main stem of a major river corridor (Skalak et al. 2013).

Nationwide, the U.S. Army Corps of Engineers reports a total of 87,359 dams;
slightly more than half of which were constructed between 1950 and 1980, with a
precipitous decline thereafter. While the majority of the dams in the United States
are low-head, earth-filled, and privately owned, most of the major rivers in the
Nation have been dammed for purposes ranging from navigation and hydroelectric power to flood control, irrigation, and recreation. Indiana has 927 dams,
most privately owned, earthen, recreational structures heavily concentrated in the
southern half of the state. Alongside their intended benefits, dams individually

produce significant changes in riparian habitat, to include converting free-flow to slack water, accelerated evaporation, erosion and deposition of sediment, water temperature, turbidity, and the mix and distribution of species. Meanwhile, the cumulative, anthropogenic impact of hundreds of dams in Indiana, tens of thousands of dams in the United States, and hundreds of thousands of dams worldwide is both significant and lightly studied.

It remains unclear whether or not the Anthropocene will officially replace the Holocene as the latest geological epoch, and simultaneously, there is on-going debate about the starting point of the Anthropocene. In a paper published in *Nature* in 2002, Crutzen argued that "the Anthropocene could be said to have started in the late eighteenth century, when analyses of air trapped in polar ice showed the beginning of growing global concentrations of carbon dioxide and methane (Crutzen 2002, 23). Writing in 2008, a distinguished group of scientists representing the Stratigraphy Commission of the Geological Society of London agreed with Crutzen on the general subject of the transition from the Holocene to the Anthropocene. They proposed and then discarded "the global spread of radioactive isotopes created by the atomic bomb tests of the 1960s" as a beginning point for the Anthropocene. These authors then concluded that for now it might be enough to pick a date, such as 1800. This, they argued, "would allow (for the present and near future) simple and unambiguous correlation of the stratigraphical and historical records and give consistent utility and meaning to this as yet informal (but increasingly used) term" (Zalasiewicz et al. 2008).

Building on that foundation invites examination of three points about the Anthropocene as it relates to the historical human interaction with rivers and with world environmental systems more generally. First, in thinking about the explanatory power that the Anthropocene has for clarifying the relationship between people and the environment, it is important to remember that ideas have a history. One of the most distinguished environmental historians practicing today is Donald Worster, who wrote the definitive study of the history of ecology, *Nature's Economy: A History of Ecological Ideas* (1977). In chapter 10, which engages the history of the science of ecology, Wooster opens with the following sentence: "In the beginning was the Word." He uses this biblical reference to highlight the fact that Ernst Haeckel coined the word *ecology*—originally *Oecologie*—from two Greek roots meaning "house" or "household" and "the study of." Worster goes on to say that "long before there was a word there was an evolving point of view, and the word came well after—not before the fact." Haeckel himself recognized his intellectual debt to this "evolving point of view" when he described the term he created as "the body of knowledge concerning the economy of nature[;] . . . the study of those complex interrelationships referred to by Darwin as the condition of the struggle for existence" (Worster 1977, 191–92).

So it is with Anthropocene. As it takes on meaning, it is important to study and understand the evolving points of view that gave rise to the word. The concept of the Anthropocene did not come out of nowhere. The reason that it caught on so quickly was because it brought into focus ideas and perspectives that had already begun to emerge in a number of disciplines. For well over a century scholars have wrestled with the idea and the significance of human beings transforming the natural world.

George Perkins Marsh—lawyer, diplomat, and conservationist—published a seminal work, *Man and Nature; or Physical Geography as Modified by Human Action,* in 1864. His preface articulates a perspective that embraces the entire earth and the central role of human agency. While his point of view was out of step with mainstream thinking of his own time, it retains remarkable resonance in the present. Marsh explained in his preface:

> The object of the present volume is: to indicate the character and, approximately, the extent of the changes produced by human action in the physical conditions of the globe we inhabit; to point out the dangers of impudence and the necessity of caution in all operations which, on a large scale, interfere with the spontaneous arrangements of the organic or the inorganic world; to suggest the possibility and the importance of the restoration of disturbed harmonies and the material improvement of waste and exhausted regions; and, incidentally, to illustrate the doctrine, that man is, in both kind and degree, a power of higher order than any of the other forms of animated life, which, like him are nourished at the table of bounteous nature. (Marsh 1864, iii)

Man and Nature contains major chapters on plants and animals (what he calls "Vegetable and Animal Species"), woods, waters, and sands.

Several twentieth-century scholars highlighted the role of people in transforming the natural world. The geographer Gilbert White defended an extraordinarily influential dissertation titled "Human Adjustment to Floods: A Geographical Approach to the Flood Problem in the United States" in 1942. White examined the interaction between human agency and flooding, explaining, "Floods are 'acts of God,' but flood losses are largely acts of man. Human encroachment upon the flood plains of rivers accounts for the high annual toll of flood losses" (White 1942, 2).[5] The marine biologist and popular writer Rachel Carson observed in *Silent Spring* (1962) that "only within the moment of time represented by the present century has one species—man—acquired significant power to alter the nature of this world. During the past quarter century this power has not only increased to one of disturbing magnitude but it has changed in character" (5–6). The French-born American microbiologist René Dubois published *The Wooing of Earth: New Perspectives on Man's Use of Nature* (1980), in which he included a chapter titled "Humanization of the Earth" (Dubos 1980). The historian Richard White (1966) published a brilliant and provocative history of the Columbia River in 1995, in which he characterized the river and its drainage as an "organic machine"—an interconnected and interdependent system composed of natural and artificial elements. White's Columbia River is as much a human-created, "cyborg-like" machine as it is a natural system.

Particularly insightful in terms of framing the relationship between people and nature is the definition of material culture developed by the archaeologist James Deetz: "that segment of man's physical environment purposely transformed by him according to culturally dictated plans" (quoted in Schlereth 1989, 294). Although Deetz definition rarely broke free of use by social scientists and humanists, its nuanced treatment of the role of human culture in making and remaking the environment is nearly as sophisticated in its explanatory power as the Anthropocene. Viewed through the definitional lens provided by Deetz, the global environment and its component parts including rivers are as much human artifacts as they are natural systems. They are, in fact, physical manifestations of human beings acting over time on the values and attitudes that form the bedrock of culture.

The second point about the Anthropocene is that it highlights the role of human agency and human culture in reshaping the natural world. It removes what has increasingly become an artificial dividing line between the natural and the cultural environment and between natural and human history. The American Fisheries Society recognized this perspective when it published *Historical Changes in Large River Fish Assemblages of the Americas* in 2005. The society's description of the volume reads as follows:

> Dramatic changes have occurred in the functioning of larger rivers because of social values and policies, land use, in channel causes, and alien species. These changes have resulted in the reduction in range and abundance of many native fish species. This book describes the historical changes observed in the fish assemblages of 27 large rivers in North, Central, and South America. (Rinne 2005)

By noting the important links between values and policies and "dramatic changes in the functioning of larger rivers," the American Fisheries Society recognized the essential role played by culture in transforming large floodplain rivers—and by extension the broader humanized environment.

The environmental history of the Great Lakes in the early 1970s provides a useful example of the interplay between science, policy, and culture. Richard Nixon was president (1969–74) during the height of the environmental movement that rested on a popular understanding of the science of ecology. The Nixon presidential papers make it clear beyond a shadow of doubt that Richard Nixon was no environmentalist; yet he signed several pieces of landmark federal environmental legislation, including the National Environmental Policy Act (1970), the Clean Air Act (1970), and significant amendments to the Water Quality Act (1972). He also created the Environmental Protection Agency (EPA) by executive order in 1970 and appointed William Ruckelshaus as its first administrator.

It is worth noting that passage of the National Environmental Policy Act, the Clean Water Act, and President's Nixon's Executive Order Creating the Environmental

Protection Agency were all pushed along by public outcry when the badly polluted Cuyahoga River caught fire in June 1969. The Cuyahoga is a tributary of Lake Erie near Cleveland, Ohio. A month later, in July 1969, *Time* magazine published an iconic image of the burning Cuyahoga that "fired" the public imagination. Unfortunately, *Time*'s fact checkers tripped up, and the magazine actually published a picture of a much more serious fire on the Cuyahoga in 1952. Despite that error, the story of the burning river and the image on the cover of the magazine augmented popular support for the environmental movement and reinforced a growing public constituency for federal action (Rotman n.d.).

Early in 1971, Ruckelshaus attempted to gain President Nixon's support for an accelerated cleanup of the Great Lakes. Ruckelshaus forwarded to the White House his cleanup plan along with a cover memo, in which he laid out arguments intended to persuade Nixon to support EPA's plan. Ruckelshaus told the president that his reputation "as a strong advocate for environmental improvement had suffered," because among other things "the very people RMN appeals to are also vitally interested in the environment. The white middle class suburbanite (particularly women) are very concerned over this issue." He pointed out that these suburbanites likely would not vote for someone they believed insensitive to the environment. Ruckelshaus added that "the one area that stands out for the environment and its degradation in the minds of the American people is the Great Lakes." Ruckelshaus then listed the eight states that touched the Great Lakes and reminded Nixon that he had won only four of those states in the last presidential election (1968).[6] The EPA administrator's message to the president was crystal clear: an effective politician who wants to win elections will pay attention to the environmental attitudes and values of the voters. A few months later, in April 1972, Nixon traveled to Ottawa, Canada, to sign the Great Lakes Water Quality Agreement. Again, it is absolutely clear from the records that President Nixon signed this agreement because he understood the power of the popular environmental movement in the United States and Canada (Scarpino 2010).

Culture also plays a powerful role in shaping the interaction between people and the environment in the present. Science alone is not enough to either understand or alter the behaviors that drive environmental change, especially when both the problems and the potential solutions are complex and the relationships between cause and effect are indirect and long term. Climate change offers a case in point. Under the headline, "In U.S., Most Do Not See Global Warming as Serious Threat," Gallup provided a March 13, 2014, update on Americans' attitudes toward climate change. In 1998, 65 percent of respondents to a telephone poll told Gallup's pollsters that they believed "global warming" was either under way or would happen during their lifetimes; the percentage of respondents who shared that point of view rose slowly to 75 percent in 2008 and then slipped back to 65 percent in 2014. Respondents who reported that they believed "global warming" represented a serious threat to their way of life stood at 25 percent in 1998, climbed to 40 percent in 2008, and then slid to 36 percent in 2014. While approximately 65 percent of Americans accepted the reality of global warming, about the same percentage also believed it did not represent a serious threat to their way of life. According to Gallup's survey in 2014, political party affiliation was a key variable in determining opinions of respondents on global warming. On the one hand, 73 percent of Democrats stated that they believed global warming had already begun, and 56 percent thought it represented a serious threat to their way of life. On the other hand, just 36 percent of Republicans believed global warming had already begun, and only 19 percent thought it represented a serious threat to their way of life (Jones 2014).

In 2014, the Intergovernmental Panel on Climate Change published its Fifth Assessment Report, which rested on the input of thousands of scientists worldwide, unequivocally stating that climate change was real, under way, and a result of anthropogenic activity. Among its many summary findings the report concluded, "Science now shows with 95 percent certainty that human activity is the dominant cause of observed warming since the mid-20th century."[7] During a time when the accumulating weight of scientific opinion demonstrated the veracity of climate change beyond any reasonable doubt, the percentage of Americans who believed that to be the case did not change, and political party affiliation was one of the most important variables in predicting attitudes toward "global warming." This situation stands in sharp contrast to the broad public constituency for "the environment" and cleaning up the Great Lakes that persuaded Republican resident Richard Nixon to sign the National Environmental Policy Act, to create the Environmental Protection Agency by executive order, and to sign the Great Lakes Water Quality Agreement. Or, for that matter, the popularized understanding of a clear association between chlorofluorocarbons (CFCs) and human health that pushed forty-six nations, including the United States, to sign the Montreal Protocol and to agree to the banning of CFCs. When it comes to changing behaviors that adversely impact the environment, what people think and believe is just about as important as the verifiable results of scientific research.

Scholars interested in using the concept of the Anthropocene need to realize that human agency is not a single, undifferentiated variable. Culture differs from group to group, and cultures evolve over time. If we are really going to understand human impact on rivers worldwide—or human impact on global environmental systems—then we need to study and understand the historical fabric of cultural contexts that produced those changes. We also need to pay attention to the unintended and unanticipated consequences of human actions.

A final point about the Anthropocene relates to the opportunity for, and importance of, interdisciplinary collaboration. When considering the history of human interaction with the environment, there are two tremendous intellectual watersheds in the past two centuries: Darwin's ideas on natural selection and the science of ecology. Both fundamentally changed the way people thought about their relationship with the natural world. The concept of the Anthropocene has the potential to become a third great intellectual watershed. If we accept the idea of the Anthropocene as an epoch in which human agency represents the most significant variable driving environmental changes on earth, then understanding those changes will take the combined and integrated efforts of scholars in science, social science, and the humanities. Part of the "magic" of the Anthropocene may be its potential for drawing scholars out of disciplinary silos and into collaborative research aimed at creating not only new knowledge but also a new synthesis that views barriers between natural history and human history as highly permeable. But then the question should become, knowledge to what end? Persuading people of the seriousness of climate change or a range of issues surrounding freshwater will require education, effective leadership, and informed policy. By itself good science will not be enough.

In the past few decades, it has become increasingly clear that human beings have done exactly what science fiction writers like Robert Heinlein and Arthur C. Clarke and Isaac Asimov wrote about in the 1950s and thereafter. That "Blue Marble" hanging in the vastness of space turned out to be our own terraformed world. Acting on the values and attitudes embedded in our cultures and employing science and technology and energy, we have literally terraformed our own planet, including hydrologic and atmospheric systems. We did not do it in the planned, ordered, science-based manner imagined by science fiction writers, and in many cases what we did not mean to do has played as much of a role as what we actually set out to accomplish.

Transformation cut in at least two directions: on the one hand, reorganized earth systems favor human beings and support a vast and growing worldwide human population; on the other, the unintended and unanticipated consequences of those reorganized earth systems pose serious threats to human societies and to the integrity of earth's environment. Among those threats are availability and distribution of freshwater, species extinction, and climate change.

A historian should be very cautious about claiming lessons from history. With that caveat in mind, a few general lessons emerge from studying the historical interplay between people and rivers—and people and the larger global environment. (1) Rarely do people set out to intentionally inflict damage. They almost always modify their surroundings for what they believe are socially beneficial purposes. (2) There are always unintended and unanticipated consequences associated with human actions. In order to really understand the Anthropocene, we need to consider what people set out to do, as well as what they did not mean to do or what they didn't see coming. (3) If the Anthropocene is distinguished by global environmental impact so far reaching that it left a stratigraphic record, then gaining insight into the emergence and evolution of the Anthropocene requires careful study of human actions driven by attitudes and values embedded in culture. In order to gain insight into what people did in the past, how they act in the present, and what they are likely to do in the future, it is essential to pay attention to the complex and subtle tapestry of culture over time.

Finally, knowledge of how profoundly past human actions transformed earth systems should go hand-in-hand with a sense of responsibility for the consequences of terraforming our own world. Writing in 1864, George Perkins Marsh saw the significance of the historical transformation of earth by human action, and the responsibility that came with that knowledge. Marsh pointed out

> the dangers of impudence and the necessity of caution in all operations which, on a large scale, interfere with the spontaneous arrangements of the organic or the inorganic world; to suggest the possibility and the importance of the restoration of disturbed harmonies and the material improvement of waste and exhausted regions; and, incidentally, to illustrate the doctrine, that man is, in both kind and degree, a power of higher order than any of the other forms of animated life, which, like him are nourished at the table of bounteous nature. (Marsh 1864, iii)

Despite our power, human beings, like all other life on earth, "are nourished at the table of bounteous nature." In the end, an obligation to be stewards working to restore "disturbed harmonies" may be the most important lesson derived from studying the history of the Anthropocene. It is a lesson that carries on its shoulders the knowledge of earth systems produced by science; insights into human culture and motivation gained from history and other disciplines; and political and policy issues that highlight the value of applied, transdisciplinary research.

NOTES

1. In 1955, the domestic American fleet of cars and light trucks averaged 3,562 pounds "curb weight" and 16 miles per gallon. Average miles per gallon of American-made cars and light trucks had fallen to a post–World War II low of 12.2 miles per gallon in 1973, corresponding with their highest postwar average curb weight of 4,022 pounds. Figures on weight and miles per gallon: http://www.nhtsa.gov/cars/rules/cafe/historicalcarfleet.htm.

2. Elkins (1999) explains the safety risk of refrigerants used before chlorofluorocarbons as follows: "Refrigerators in the late 1800s and early 1900s used the toxic gases, ammonia (NH_3), methyl chloride (CH_3Cl), and sulfur dioxide (SO_2), as refrigerants. After a series of fatal accidents in the 1920s when methyl chloride leaked out of refrigerators, a search for a less toxic replacement began as a collaborative effort of three American corporations—Frigidaire, General Motors, and Du Pont."

3. When Rachel Carson published *Silent Spring* in 1962, concerns she helped to highlight—the widespread, indiscriminate use of synthetic, organic pesticides and related chemicals—jump-started the modern, ecology-based environmental movement.

4. For a relatively recent and cautious examination of the possibility of a sixth mass extinction, see Barnosky et al. 2011. This article also contains a thorough bibliography. Also helpful for understanding the science and the assumptions underlying examination of a possible sixth period of mass extinction is De Vos et al. 2015. The author thanks Dr. Samuel Scarpino and his colleagues at the Santa Fe Institute for recommendations on literature related to mass extinction.

5. See also Scarpino 1997. The relatively recent work by Hamilton and Grinevald (2015) offers a useful discussion of writers (largely scientists) who called attention to the profound human impact on

the natural world and who many recent writers have identified as having foreseen the idea of the Anthropocene. Hamilton and Grinevald argue that it has become "accepted wisdom that the Anthropocene was foreseen by scientists in the 19th and early 20th centuries," and although "the present authors initially accepted this view, after critical reflection and rereading the historical sources we now disagree with this intellectual phylogeny" (60). While this article seems to confuse the history of an idea with "foreseen," it nonetheless offers a highly useful overview and a very helpful bibliography.

6. William D. Ruckelshaus, Administrator, EPA, to John C. Whitaker, "The President and the Environment," 11 January 1971, National Archives and Records Administration II, College Park, MD, Nixon/Whitaker, Box 135, Great Lakes Agreement, 2 of 2. Cited in Scarpino 2010.

7. See, e.g., the foreword to *Climate Change 2013: The Physical Science Basis*, the Working Group One Contribution to the Fifth Assessment Report of the Intergovernmental Panel on Climate Change, which states, "*Climate Change 2013: The Physical Science Basis* presents clear and robust conclusions in a global assessment of climate change science not the least of which is that the science now shows with 95 percent certainty that human activity is the dominant cause of observed warming since the mid-20th century." See also Intergovernmental Panel on Climate Change, *Climate Change 2014: Synthesis Report Summary for Policy Makers*, 4: "Anthropogenic greenhouse gas emissions have increased since the preindustrial era, driven largely by economic and population growth, and are now higher than ever. This has led to atmospheric concentrations of carbon dioxide, methane and nitrous oxide that are unprecedented in at least the last 800,000 years. Their effects, together with those of other anthropogenic drivers, have been detected throughout the climate system and are extremely likely to have been the dominant cause of the observed warming since the mid-20th century." Also: "About half of the anthropogenic CO_2 emissions between 1750 and 2011 have occurred in the last 40 years (high confidence)." The IPCC was created in 1988. It was set up by the World Meteorological Organization (WMO) and the United Nations Environment Programme (UNEP) as an effort by the United Nations to provide the governments of the world with a clear scientific view of what is happening to the world's climate.

PART THREE

Experiences

The chapters that follow consider the place of *anthropoi* in the Anthropocene. In the face of unprecedented human-environment interactions, how will communities engage with the challenges of living in a bioengineered world? There is already a strong narrative among Anthropocene scientists that humanity should resist despair in the face of unprecedented challenges such as global climate change. Instead, the path lies open for us to imagine communities that not only survive but also thrive in this new epoch. The studies in part 3 offer examples of the resilience and adaptability of human societies, which, over the past 250 years, have responded creatively to the challenges of the Anthropocene. The "experiences" examined here interrogate the capability of human communities past and present to respond to moments of fracture and crisis.

In the late 1700s, at the dawn of the modern European Industrial Revolution, state mechanisms were relatively weak, requiring people to respond locally to specific problems. Using the example of a flood in eighteenth-century Newcastle, Berry's chapter 9 shows how community self-organization anticipated the ways, if not the means, by which grassroots environmental activism would later organize to lobby for action in the face of political intransigence. By the twentieth century, state mechanisms became formalized, and new forms of artificial boundaries were raised between competing jurisdictions and commercial interests. Livelihoods that were sustained in traditional river cultures were replaced by industrial-scale exploitation of riverine resources by multinational corporations (consider the value of hydroelectric energy, fisheries, and river transportation). Likewise, communities in both industrial nations and the developing world faced new challenges from urbanization and population pressure. For a time, many rivers that are at the heart of the world's major cities were forgotten, becoming open sewers that were too poisonous to harbor life.

While Berry's chapter considers a historical experience of environmental devastation—and how one community organized itself in response—the remaining chapters examine the contemporary context and how people live with their Anthropocene riverscapes. Kane focuses on how changing approaches to geoengineering in Singapore shape and reshape communities and cultural practices. Using the idea of "front- and backstage urban transformations" Kane shows how anthropology can uncover the constantly shifting interactions between society, culture, and environment. Miss and Carter provide a case study of environmental public art practice. Using several installations in Indianapolis, Indiana, they show the importance of artistic interventions in environmental consciousness. The book ends with a reflection by Matt Edgeworth on the Chicago River, a striking example of a river of the Anthropocene. Using a canoe to experience the river, Edgeworth takes the reader on a phenomenological journey to explore it as a "hyperobject."

The chapters in this part offer some hope that human societies have the capacity to co-create a more sustainable future that acknowledges the finite quality of our natural resources but only if the idea of the "commons" prevails over narrower concerns of commercial profit and short-term gain. Flourishing communities of the future will have acknowledged that watersheds, floodplains, and confluences do not respect political boundaries. More than gross domestic product, biodiversity and human well-being are better measures of health in coupled human-environment systems, of which rivers are our prime examples.

9

The Great Tyne Flood of 1771

Community Responses to an Environmental Crisis in the Early Anthropocene

Helen Berry

The Anthropocene presents humanity with environmental challenges on an unprecedented scale that can seem unfathomable and daunting. Scientists have debated the big data that measure the impact of the "Great Acceleration" (Steffen 2015) on earth systems (such as rising pollution and sea levels, sinking deltas, and severe weather events linked to climate change). Social scientists and humanities researchers are examining the implications of these changes for societies across the globe, from economists who address growing inequalities in the distribution of wealth to political theorists and legal experts who question whether current mechanisms for national and international governance are fit for these radically altered times (unlike politicians and bureaucrats, hurricanes do not respect geopolitical boundaries; neither do river catchments under flood conditions).

Part of the process of trying to make sense of complex and deeply linked environmental, economic, and social change in the twenty-first century has been the attempt to find precedents and strategies for survival by looking backward as well as forward in time. The "microhistorical" approach is a widely used methodology in historical research and is an attempt to reconstruct a particular historical moment in context—often through the selection of a moment of disruptive change such as a riot, a show trial, or a transferral of power from one person or body to another (Lepore 2001). The microhistory offered in this chapter explores the ways in which the local population in the Tyne River valley in Northeast England responded to one of the most catastrophic natural disasters in its modern history—a flood that took place on the night of November 16–17, 1771. It is based on previously undiscovered archival evidence that came to light in the summer of 2013 in the archive held by the Society of Antiquaries of Newcastle—a previously uncataloged book of documents and claims relating to the organization of compensation for flood relief victims (hereafter SANT/BEQ). The chapter starts with a brief account of this flood and the extent of the damage it caused, then turns to consider how a disaster relief committee was organized at very short notice as well as the methods that they devised for compensating flood victims. The inadequate mechanisms of local government coupled with an emerging nation-state without a national task force for dealing with environmental disasters required an innovative and swift response from people with the social rank, authority, experience, and resources to provide relief in the absence of alternative power structures. Some of the issues and challenges faced by those who were flood victims, and by those who tried to restore both transport infrastructure and economic and social stability—not to mention the safety and well-being of those affected—provide a case for comparison with social responses to other flood crises in different time periods and riparian cultures (e.g., Welky 2011). As such, it explores the opportunities and constraints faced by a proto-industrial society in the face of an environmental catastrophe. Finally, some general remarks are made by way of conclusion about the potential for historians' storytelling to engage wider audiences and motivate communities to engage with education, conservation, and policy formation by raising awareness of local river cultures.

The particular example of an eighteenth-century flood event in Northeast England merits consideration amid the uncertainties of our present circumstances. Some of the most influential contemporary thinkers whose work has transcended narrow disciplinary boundaries have embraced Churchill's formulation that "the farther back you can look, the farther forward you are likely to see" (Guldi and Armitage 2014). The historians Jo Guldi and David Armitage, responding to "big data" on climate change, argue that "renewing the connection between past and future, and using the past to think critically about what is to come, are the tools that we need now. Historians are those best able to supply them" (2014, 13). Yet, as these historians point out, it was scientists who first became embroiled in what was essentially "a controversy about history," a "major public battle" over the chronology and character of the Anthropocene, initiated by the Nobel Prize–winning chemist Paul J. Crutzen (Crutzen 2002; Crutzen and Steffen 2003), that became the primary task of the Anthropocene Working Group (Syvitski 2016). In the search for more effective and sustainable solutions to earth systems governance, scientists have looked to human history to provide models for government and market economies whose footprints (however defined) on ecosystems were light. Here historians could offer an as yet unrealized potential to contribute to the project of creating a sustainable future. Historians of different time periods and cultures have the knowledge of diverse precedents that scientists are seeking. We are also good at gathering and sifting evidence that can be transformed into meaningful narratives that help to make sense of big data not only for academic audiences, but the wider public. We know how to interrogate causality and address the impact of continuity and change over time. Usually this is not the "deep time" of prehistoric geological eras but the relatively short time frame of human history recorded in language, for the sake of argument the past ten thousand years (Corfield 2007).

Unlike geologists, or indeed our closer colleagues in archaeology, most historians work within the much narrower prescriptions of one or two centuries of expertise, although how time frames are divided is mostly a culturally specific as well as discipline-specific determinant. Guldi and Armitage (2014) assert that historians must return to analyzing longer time frames because of the pressing need to consider "big data" and broader processes of change over time presented by environmental history. This has more often been the case in economic history, where researchers have mapped and quantified the transition from an organic preindustrial economy in the West (reliant on wood or charcoal for power generation) to one based on fossil fuels. A recent, innovative example is a highly influential book on energy in the Industrial Revolution by one of the most influential contemporary scholars on the subject, E. A. Wrigley. Wrigley reconceptualizes economic change through the long-term environmental shift from direct (organic) reliance on plant photosynthesis to new production horizons fueled by coal (Wrigley 2010, 14). Societies built on the organic economy, he observes, are consigned to what the classic economic historian W. S. Jevons called "laborious poverty," whereas surplus wealth and the rise of tertiary sectors of the economy flow from the exploitation of fossil fuel (Jevons 1906). Some accounts of industrialization are embracing environmental history, yet it is still not uncommon to find analyses of the transition to modern society based on fossil fuels that paint a broadly positive picture of human progress. Economic histories of the English Industrial Revolution tend not to address the environmental impact of these processes and have yet to address their contribution to the Anthropocene directly. By contrast, elsewhere there are examples of histories that integrate the environmental consequences of mineral exploitation and river engineering (Scarpino 2014) and those that have charted the collapse of civilizations built on finite resources (Diamond 2005; Davies 2012).

Reevaluating the eighteenth and nineteenth centuries and Western European industrial history within the conceptual framework of the Anthropocene presents challenging, even revolutionary, possibilities for a totally new critical framework. The present chapter asserts the value of microhistory as a powerful vehicle for forensic analysis of disparate forms of evidence, as well as the creation of meaningful narratives around key issues that are commonly witnessed in the Anthropocene. Assuming, for the sake of argument, that we accept the original hypothesis propounded by Crutzen and Steffen that the Anthropocene began around 1800 (Crutzen and Steffen 2003, 254), the critical phase of the "early Anthropocene," marked by the rise of fossil fuel exploitation and rapid urban development in the West, ought to merit detailed historical reevaluation. The conditions we are living with today—increasingly frequent flood events, changes to weather systems, rising sea levels, and the rapid disappearance of sea ice—are the accumulated

FIGURE 9.1. John Hilbert. Medieval Bridge, Newcastle upon Tyne, ca. 1727. Engraving. By permission, Newcastle City Library (accession no. 15399).

consequences of industrialization processes that began to develop rapidly in the late 1700s. The birthplace of the world's "first industrial nation," Northeast England offers a case study of how the processes of industrialization quickly diversified, replicated, and refined elsewhere in Europe and on the North American continent (Crosby [1986] 2004). In the nineteenth and twentieth centuries, industrial transformations based on fossil fuel exploitation were witnessed globally, from the Indian subcontinent to the Far East and China to Latin America (Osterhammel 2014). What happened in Newcastle during a sudden flood event under conditions of the early Anthropocene could provide clues about the long-term trajectory of the industrialized world.

· · ·

The first days of November 1771 were marked by incessant rain and northeastly winds. To the northwest, near the source of the Tyne past Corbridge, the Solway Moss bog became saturated and flooded the rich farmlands populated with livestock (Donald 1774). The harvest of oats, a local crop, and hay for overwintering sheep and cattle was already gathered, but the water seeped into storage barns and ruined precious crops. To the south, the tributaries of the Tyne swelled into a raging torrent that by 11:00 P.M. on November 16 had raged down the valley, gathering speed and sweeping everything away—crops, cattle and people, even buildings. In Newcastle, the five-hundred-year-old medieval bridge (fig. 9.1), sorely in need of repair, began to creak and topple. A bottleneck was created between the piers of this ancient structure by silting—a problem exacerbated but not entirely caused by ships ballast-dumping sand that was not solved by the regular attempts to remove as much as 100,000 tons a year by dredging. Very little about the Tyne was "natural" in the eighteenth century; human intervention in river systems in Europe and across continents can be traced to prehistoric times (Edgeworth 2011). In England, there was an acceleration in the rate of river management during the medieval period (White 1962). From at least the fifteenth century, the Tyne was dammed, fished, and used as a source of water power and the site of industrial production. Its banks were farmed and agricultural waste and silt ran off into the river via tributaries from the Upper Tyne to the confluence with the North Sea (Wright 2014). The source of the Tyne in upland areas with sparse vegetation and rough terrain was lightly populated and rural in character. It gave rise to only one town of significant size, Newcastle, which had a population of thirty thousand people by 1700. In the 1600s, there was already significant lead mining activity in the Upper Tyne region. By the early 1700s, three quarters of a century before the period usually associated with the Industrial Revolution, the Tyne was already a working river, used to transport coal from open-cast mines via flat-bottomed boats (or keels) to the collier ships anchored off Tynemouth, ready for transportation to London.

Many schemes had been devised to solve the problem of silting on the Tyne, mainly a human-induced problem that hindered the commercial life of the river, but there existed ancient and conflicting interests that mitigated against a joined-up solution to the problem. The problem was made worse by structural engineering: the old Tyne bridge, situated about 8 miles from the mouth of the river, further encouraged silting. This bridge was on the approximate site of one dating back to Roman times, and it had a practical and political function. The only crossing point for human traffic and goods for several miles, its apex marked the point between two jurisdictions—on the Newcastle side, the rights of the incorporated Newcastle Council (whose powers were granted by Royal Charter) and the quasi-feudal jurisdiction of the bishop of Durham, whose rights extended over the Gateshead (southern) banks of the Tyne in County Durham and who had the power to levy charges for maintenance but preferred instead to divert money to the coffers of the church. Repairs were haphazard, and as figure 9.1 indicates, the people living, trading, and traveling on the bridge did so at their own peril. "Pontage"—an ancient tax on using the bridge by the local guildsmen, such as fullers, dyers, glaziers, goldsmiths, and weavers—was collected erratically, and royal grants were erratic. The famous engineer John Smeaton, who was known for building lighthouses in

To the Right Worshipful John Erasmus Blackett Esq. Mayor of Newcastle upon Tyne. This VIEW of the RUINS of the BRIDGE of that TOWN, as they appeared after the Fall thereof in November 1771, Is most respectfully Inscribed by his very obliged and most devoted faithful humble Servant John Brand. October 7th 1772

FIGURE 9.2. Engraving showing postflood ruin of the Tyne Bridge. Illustration from John Brand, *History and Antiquities of the Town and Country of Newcastle upon Tyne,* 2 vols. (1789). By permission, Dr. Peter Wright (private collection).

the southwest of England, was called in to do a survey of the bridge, and estimated in the year before the flood, 1770, that £150–200 were needed for urgent repairs—but nothing was done (Garret [1818] 2010).

As the small hours of the morning of November 17, 1771, wore on, and the water surged from higher up the Tyne valley, people at the quayside in Newcastle began to evacuate their dwellings and flee for their lives. Houses began crashing into the river. A catastrophic sight greeted the townsfolk of Newcastle and Gateshead, north and south of the River Tyne on that morning (fig. 9.2). The river had risen eight feet above the high-water mark of an average spring tide. Dwellings in the Sandhill area just along the quayside were six feet underwater. Coal ships had been lifted onto the quay. Keel boats, debris, and timber littered the riverbanks. Two of the twelve low stone arches of the bridge were swept away. One distraught witness, a Mrs. Fiddas, witnessed one of the arches collapse and carry away her husband and a maid. There were other fatalities: Byerley the ironmonger and his son, Ann Tinkler, a draper, and an apprentice to James the cheesemonger. Many bodies were never recovered. In one account, the strange sight of one of the houses that had been on the bridge belonging to Patten the draper, floated down the river, and the local newspaper reported that in it were an unharmed dog and cat. By 4:00 P.M., the floodwaters had subsided.

The fall of the Tyne bridge had social and economic ramifications of national and international importance. North and south were effectively cut off from road communication, and the maritime coal trade was disrupted by the flooding of wagonways that transported coal from mineheads to the loading points for the river "keels" that transferred their cargo to collier ships. The only other crossing point on the Tyne, farther upstream at Corbridge, was a seventeenth-century stone construction—a pack bridge for sheep dating back to 1674—and this survived. There were twenty-five recorded fatalities (Northumberland County Council 2010) and hundreds of families displaced from their homes. Formerly affluent households were said to be reduced "to the most abject misery and want," with some of the poorest folk left with nothing but the clothes they were wearing when they abandoned their homes (*Narrative of the Great Flood* 1772, 2–3).

It is difficult to separate out the enlightened self-interest of the ruling elites in the eighteenth century from their philanthropic and charitable activities—but these, rather than strategic government initiatives, were the only source of solutions to the problems caused by this particular environmental catastrophe. It is here that the newly discovered archive of flood disaster-related documents provides invaluable information about what happened next. Opening a subscription book was a common method for raising donations for a cause in the Georgian era, and it was a familiar mode of organization for the ruling elites to adopt in response to a crisis. Money was raised in this way for one-off charitable causes, capital building projects, and charitable institutions such as hospitals (Butler 2012). Lists of donors were printed in the hierarchical order of social precedent in English society (fig. 9.3), usually starting with the nobility and ruling elite and proceeding through the ranks of professional men and local councillors. Within just a few weeks, churchmen, titled families, the Corporation of Newcastle, donors anonymous and named, from as far afield as Scotland and London, started to send donations to the relief fund. Newcastle had a precociously developed print media at this time—it was one of the earliest provincial towns to have a newspaper—and the pages of the *Newcastle Courant* kept readers updated about the consequences of the flood and how to donate to assist victims.

The subscription list bypassed the usual method for providing poor relief, which at this time was administered at the parish level, reflecting the fact that relief was provided by one-off charitable donation rather than a local levy. The flood, of course, did not respect parish boundaries; there were in fact at least fifteen historic parishes flanking the Tyne that had the potential to be affected by the floods (figs. 9.4, 9.5), although those in Newcastle and north of Newcastle were more badly affected; parishes downriver of Newcastle such as South and North Shields were protected by the presence of "Jarrow Slake"—a bend in the river

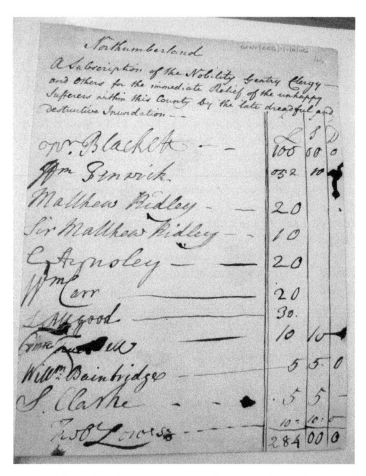

FIGURE 9.3. "A Subscription of the Nobility Gentry Clergy and others." The start of the list of flood relief donors in the county of Northumberland, showing amounts donated (1771/2). SANT/BEQ/1/1/4/46. By permission, Newcastle Society of Antiquaries.

enhanced by a man-made culvert—which helped to direct water away from habitation and farmland. Other river systems were affected by the flooding from the same severe weather events, giving rise to simultaneous crises across neighboring counties to the south and west, specifically County Durham, north Yorkshire, and Westmoreland.

ORGANIZATION AND SOCIAL ACTION

The first meeting of what became the disaster relief committee was in the market town of Hexham on December 19, 1771. The committee's first resolution was

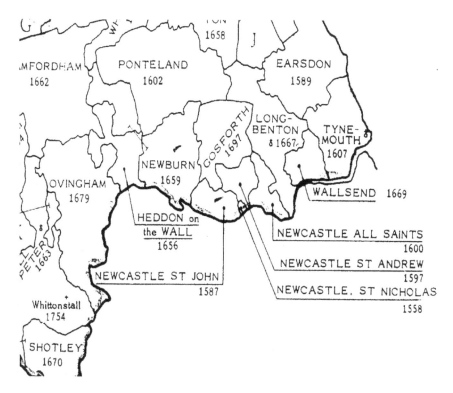

FIGURE 9.4. Historic parishes of the Tyne (a) North of the Tyne. By permission, Dr. Peter Wright.

that subscription books were to be opened for charitable donations in Newcastle, Hexham, (South and North) Shields, Morpeth, Alnwick, Belford, Rothbury, Wooler, Stamfordham, Bellingham, "Haltwezel" (Haltwhistle), and Berwick. The second resolution was that subscribers should pay their money upon subscription as humanitarian need was urgent ("the objects of this Charity are in Want of immediate Support"). The committee comprised male vested interests and the propertied elite—"33 Gentlemen and Clergymen residing near to the River Tyne where the principal damage occurred"—who were appointed "to distribute the money, assess individual loss suffered, their present condition and circumstance, and calculate the distribution of money accordingly" (SANT/BEQ/1/1/4/1–2). Women did not number among the committee, although they featured prominently among donors to the charitable relief of flood victims. The committee followed through with the decision to publicize their activities in the local press. At their ninth and final meeting (July 6, 1772), "The Committee having made a final Distribution of the Subscriptions it is Ordered that the Secretary do send to the

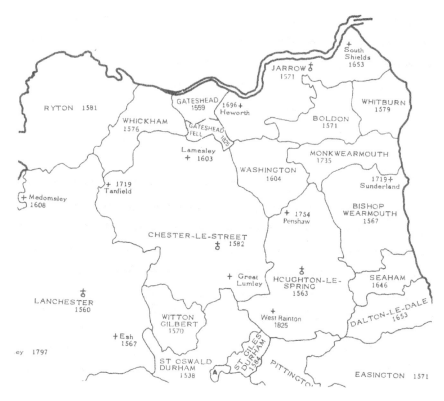

FIGURE 9.5. Historic parishes of the Tyne (b) South of the Tyne. By permission, Dr. Peter Wright.

printers of the Newcastle Newspapers a General State of the Account to be by them inserted in their Papers."

The ninth resolution limited the time frame for the existence of this charity, setting finite goals for its operations: the subscription books were to be open until January 20, 1772, and no longer. The eleventh resolution was that once all money had been distributed, the committee resolved to publish in the Newcastle newspapers "an account of their Receipts and Disbursements." The names of the thirty-three members of the committee were listed and published. Meticulous record keeping characterized the committee's operation. What is remarkable is that in the context of England at this time the legal and bureaucratic mechanisms and infrastructure existed to organize relief with relative speed. Essential features were bureaucratic probity, account keeping, and respect for the exercise of trust on behalf of professionals charged with this responsibility, which helped to facilitate the allocation of resources. The press played a crucial role in raising money by subscription and communicating the activities of the disaster relief committee, in effect making the process transparent and accountable. By contrast, the Common Council in Newcastle had entrenched and vested interests that made their response slow and (many felt) negligent to the urgent needs of the local people, not just in supplying the basic needs of the people who had lost homes and possessions, but also the traders whose supply routes had been interrupted. The Corporation of Newcastle was expected—and to some extent did—assume the initiative in organizing relief and repairs to the infrastructure. Within a year of the flood, a ferry service was quickly provided for the local mail, £2,400 were set aside to build a temporary bridge, and engineers were commissioned to consider options for rebuilding a permanent structure (Newcastle Common Council Minutes 1772). It would be anachronistic to expect the council to have acted as comprehensive providers of a coordinated humanitarian relief program at this time, although they did make a collective donation to the relief fund.

The need to ration relief donations became immediately apparent to those appointed to administer charitable donations. Their response reflected an increasingly entrenched class system in English society but proved an effective (if controversial) form of triage. At their fourth meeting (February 5, 1772) the administrators discussed distributing funds among "Sufferers" (flood victims): the "first Class of Sufferers" or "distressed Sufferers" were those deemed to be in most urgent need, without a roof over their heads in many cases and little or no means of subsistence; "Second Class of Sufferers" were the less urgent cases whose livelihoods had nevertheless been affected severely; "the third Class" were those whose nonurgent claims for compensation could be deferred to a later date (fig. 9.6). Although the charity's main patrons were drawn from the most powerful ruling elites in the region, those who donated included both women and men from relatively humble backgrounds. Of the 495 individually named subscribers, only 10 are listed with their occupations. North Shields and Hexham recorded the occupation of some (male) subscribers: Hexham recorded 7 subscribers' occupations (of its 122 entrants): "Barber, Blacksmith, Butcher, Clogger, Hardwareman, Tailor, Watchman"; North Shields recorded 3 subscribers' occupations (of its 5 entrants): two attorneys and a surgeon. Some subscription lists include the titles of some entrants that denote status or rank: 2 "Sea Captains," Stamfordham and Newburn; 12 clergy from the towns of Morpeth, Wooler, and Haltwhistle and another unspecified area; two physicians across Wooler and an unknown location. Higher up the social scale was a donor who was a baronet from Stamfordham (SANT/REQ/1/1/4/17–20, 22, 28–31, 34, 35–40, 43–44, 91). From the outset, there was a marked variation in levels of donation: those in the Dilston area who had suffered few direct effects of the flooding contributed nothing "tho' rich," while others in the Corbridge area, though they were also spared the worst flooding, "have given liberally" (SANT/REQ/1/1/4/60).

Another aspect of the social organization of capital that made relief efforts more effective in this context was the development of a local banking network

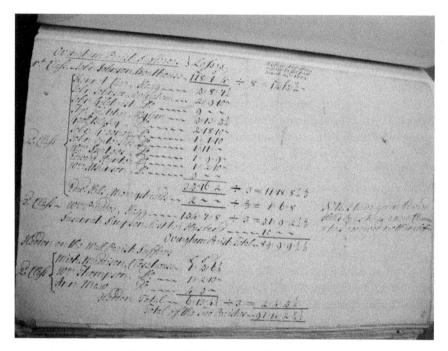

FIGURE 9.6. Categories of recipients of relief in the parishes of Ovingham and Heddon-on-the-Wall, Northumberland (showing different "classes" of sufferers, left-hand column). SANT/BEQ/1/1/4/74. By permission, Newcastle Society of Antiquaries.

and the presence of professional men who were trusted citizens known for their probity and administrative skills. Deposits of donations for the relief operation were made in two Newcastle banks, with scrupulous recording of receipts by the committees for each county. Disbursements were made according to need, in line with the principles set out by the found Subscription Committee in Newcastle. Trusted professionals volunteered their administrative skills, specifically attorneys such as Ralph Heron, one of the most active and efficient administrators of the Northumberland County donations, and clergymen who were trusted to act as loss assessors across the region.

Loss assessors signed that they had delivered "regular and just" estimates to the Subscription Committees, detailing household by household the specific goods, livestock, and crops lost, damaged, or destroyed (fig. 9.7). Damage done to buildings, land, fences, and grain sown was not included in the estimates made by loss assessors. Compensation was then paid pro rata, according to the "class" of sufferer, as categorized by the Subscription Committee overseeing the disbursement of donations for each county (tables 9.1, 9.2).

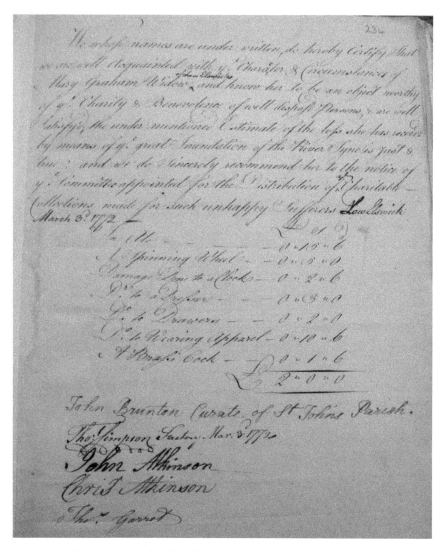

FIGURE 9.7. Loss assessment for Mary Graham, widow, of Low Elswick. March 3, 1772. SANT/
BEQ/1/1/4/f. 234. By permission, Newcastle Society of Antiquaries.

HUMAN CONSEQUENCES: CONFLICT

As noted, there was variation in the amounts donated across the region, with some
areas displaying considerable generosity and others almost none. The committee
appointed to oversee the process quickly ran into complications and conflict. One
high-profile donor, Alderman William Fenwick of Bywell, demanded his money

TABLE 9.1. Amount of compensation paid for flood damage by category of "sufferers"

	£	s.	d.
First-Class Sufferers	989	6	7
Second-Class Sufferers	415	15	5
Third-Class Sufferers	354	11	1
Persons not in any particular class	139	9	7
Expenses attending the Committee	65	10	0
Total payments made	**1,964**	**3**	**9**

SOURCE: River Tyne Flood Papers (SANT/REQ/1/1/4/13).

TABLE 9.2. Subscriptions collected January–March 1772 by geographical location

Subscription Book Totals	£	s.	d.
Northumberland	92	11	0
Hexham	12	12	6
North Shields	3	8	0
Morpeth	44	13	6
Alnwick	56	14	0
Belford	51	5	0
Wooler	79	0	6
Bellingham	7	10	6
Stamfordham	26	5	0
Berwick	146	7	6
Haltwhistle	12	3	0
Whitley in Hexhamshire	13	4	0
Newburn	9	6	0
Corbridge	15	3	0
Haydon	9	1	3
Unknown location (coll. by Rev. Allan)	40	12	0
Total	**619**	**16**	**9**

SOURCE: River Tyne Flood Papers (SANT/REQ/1/1/4/17–45 and 53–91).

NOTE: These are not total figures for amounts raised (see table 9.1) but illustrates distribution by region: there is an additional Northumberland Subscription Book (SANT/REQ/1/1/4/46–52).

back so that he could donate to specific families (SANT/REQ/1/1/4/54). A suggestion was made that the subscription lists for donations should remain open for a longer time, given the logistical difficulties of receiving and distributing money to and from counties at a greater distance from Newcastle. More seriously, there was a dispute over whether sufferers in Newcastle or the surrounding counties were benefiting disproportionately from charitable donations, with a meeting advertised in the local press to agitate for a review of whether compensation was being fairly distributed. The authority of the founding Subscription Committee was challenged, as was the legitimacy of putting all donations into Newcastle banks rather than ensuring local people were compensated more immediately from donations in their local area. At its worst, the fallout from the flood catastrophe of 1771 highlighted the preexisting tensions that existed within and between urban governance and the "handmaiden" status of its rural hinterlands in the English provinces. (SANT/BEQ/1/1/4/11–13). Though the River Tyne was already well on its way to becoming a fully human-engineered river system by the second half of the eighteenth century, the interconnectedness of cause and effect, of human action and a chain of consequences from source to sea, was not well understood in the era of proto-industrialization, characterized as it was by fragmentary governance, local and competing hierarchies of power, and divided political jurisdictions. As Jason M. Kelly highlights in the first chapter of this volume, social inequalities were reinscribed from early on in the Anthropocene, with varying degrees of suffering experienced at different levels of society in the same catastrophic flood event. Perhaps on these grounds, it is indeed valid to speak of more than one "Anthropocene": at least one for the rich, one for the poor.

. . .

This is a case study in the action taken as a response to a flood disaster in the early Anthropocene situated amid the process of coal-powered industrialization, of which Newcastle and its hinterland were the major source in the English Industrial Revolution. The extreme flood event of November 1771 on the River Tyne was preceded by several other recorded floods in the eighteenth century on the Tyne and its neighboring river systems. It was also followed by subsequent flood events in the nineteenth century, although none matched the severity of the 1771 flood, measured in terms of fatalities, disruption to the transport infrastructure, or loss of assets in the form of property and livestock (Northumberland County Council 2010). The flood thus was partially a natural disaster caused by an extreme weather event, but its disastrous effects were also the result of human modification of the River Tyne catchment, evidenced in recurrent river silting caused by agricultural development along the riverbanks and ballast dumping, industrial processes such as mining, and alterations such as calverts to the flow and course of the Tyne and its tributaries.

Attempts on the part of local and national authorities to address the crisis that followed within existing political frameworks and traditional jurisdictions were fragmentary and largely ineffective. In the context of England at the end of the eighteenth century, it was patrician values, and a paternalistic concern for the welfare of parishioners, that drove relief efforts at the local level. The Corporation of Newcastle, made up largely of coal-owning local magnates, intervened to rebuild the transport infrastructure so as to allow the resumption of the coal trade, and the road connection between north and south of the city, as quickly as possible. The rural catastrophe wrought by this flood is thrown into particularly sharp relief if we consider France in the 1770s, and the starvation and food riots that precipitated the Revolution of 1789. The River Tyne flood disaster of 1771 could have had serious national and even international political ramifications if no compensation had been forthcoming to agricultural workers and their families. The balance of economic drivers, social and political stability, remains in constant jeopardy today in the face of severe weather events, climate change, and river flooding around the world. The rhetorical gloss put upon charitable responses to environmental disaster must be regarded now, as then, with some skepticism:

> How much so ever we may unhappily be divided amongst one another in religious or political sentiments, all seemed to unite in that spirit of charity and benevolence which so remarkably characterises the English nation. (*Narrative of the Great Flood,* 1772)

Public anger in response to on-the-ground difficulties in providing timely and adequate relief was present in the early Anthropocene, and was a forewarning of the inability of modern governments to respond adequately to environmental crises of much greater magnitude.

Looking at this English environmental disaster before the rise of the modern nation-state reminds us of the importance of local responses to catastrophes with potentially global ramifications. If the nation-state fails to provide adequate solutions to these catastrophes, and patrician responses by the local ruling elites are an anachronism, then on what do we rely? Questions of scale are critical. Thinking beyond local politics and statutory agencies, going where the floodwaters go, considering the ways in which rivers transcend the artificial boundaries imposed by human interaction, must be one response. This is an approach seen already in the establishment of River Trusts in England, which is a network of not-for-profit organizations formed by volunteers and environmental specialists over the past thirty years to work with local communities to improve habitats, educate schoolchildren, lobby policy makers, and take a holistic long-term view as guardians of river catchments who oversee ongoing regeneration. Like the donors of small amounts to the flood disaster of 1771, local peoples in the future must feel invested in the solutions that are brought to their door—sometimes literally and sometimes via the media and the imagined community of mutual interests in a riverine culture that flows so often unnoticed, until disaster strikes. This chapter has proposed that historians are useful, even essential, to interpreting large and complex data and archival evidence for outward-facing public engagement purposes; for our skills are at providing interpretation and making sense of narrative. It is essential that we work toward developing a common language and framework for the environmental challenges that lie ahead.

ACKNOWLEDGMENTS

I am extremely grateful to Lindsay Allason Jones, OBE, president of the Society of Antiquaries, for bringing this remarkable archive to my attention; to Scott Ashley, Caron Newman, Philip Scarpino, and James Syvitski for additional references; and to Ria Snowdon and Peter Wright for assistance in preparing data for this chapter.

10

Engineering an Island City-State

A 3D Ethnographic Comparison of the Singapore River and Orchard Road

Stephanie C. Kane

The top layers of the earth's crust have been remade by dense infrastructural, architectural, and sculptural conglomerations through which once-wild rivers flow. Engineers mediate the material interplay of humans and rivers, building subway and sewage tunnels, reservoirs and pipe networks, sidewalks, quays, and blocks of apartment towers to enable urbanist dreams of beauty, safety, and efficiency. Earth scientists stretch cultural geographic understanding of landscapes within geological time. As an ethnographer of water infrastructure in the Anthropocene, I frame understandings of material spheres, decision making, and social action across historical and geological time. Rivers flow in and out of geological epochs into the longue durée, the enduring structures underlying the events of human history, and into present-day arenas of socioeconomic interaction. In their effects, time scales proceed simultaneously even as they also move from past to present.

By taking into account geological action, we soon recognize that the sites within which we ground knowledge production about our environments are unstable. In other words, if we are to understand and act upon the knowledge that humanity is a geological actor (and its corollary that "there is no stable point," as Doreen Massey argues), we find that we need to live and work as if we indeed *feel*, not just think, this set of existential facts (Massey 2005, 130–42). Our ongoing conversations in the transdisciplinary space of the Rivers of Anthropocene project can be part of such empirical processes of realization. The site-based specificities of time-tagged processes emerge in tandem from scholarship and from the worlds we study. The insight that the practice of art, science, engineering, history, and culture are indeed entangled encourages us to turn away from institutionalized hierarchies of knowledge production, especially the aspects most implicated in creating the conditions for the present planetary conundrum. This realization encourages us to perform an open-minded, ethical, and collective *untangling* that re-cognizes diverse and multiscalar interplays among humans, nonhumans, and elemental forces that compose life in earth, site by site.

In my ethnographic practice, I tack back and forth in time and space doing fieldwork at one or more sites in the world and reading relevant scholarly literature; deciphering and tracing the frameworks of interpretation and infrastructural interaction at play; discovering place-based insights, patterns, and processes that can be shared across transdisciplinary riverine space (Kane 2012). To capture and represent the dynamic constellation of forces, conditions, and symbolic meanings that come into play, I extend traditional fieldwork with its "thick description" of particular cultures (Geertz 1973) and work toward restoring the "social thickness" of globalized infrastructural processes in local spaces (Sassen 2006; see also Graham and McFarlane 2015). The Singapore project presented here assembles a human-made geological subject amenable to ethnographic exploration, that is, the three-dimensional, infra-structured dynamics of daily life in riverine neighborhoods. This small piece of the ethnography of Singapore offers insight into a dynamic, shifting pattern between frontstage and backstage river-human action that operates chiefly between meso- and microscale.[1] (In illustration of the spatial range between meso- and microscales consider the "Sumatra Squall," a line of interlinked thunderstorms accompanied by strong gusts of wind that come across the Malacca Strait to hit Singapore and cause localized flash flooding problems.)

As a tiny, low-lying, tropical, urbanized surface, the island city-state of Singapore is a rich experimental domain. It has been a continuously important node in the global maritime trade networks from the sailing vessels of the third century to the twenty-first century petroleum-based container shipping industry (Malay Heritage Centre 2013; Tan 2016). Freshwater provision for inhabitants and circulating traders has always been a precondition for economic survival. Singapore's contemporary efforts to assure sufficient supplies while mitigating flash floods have led to extraordinary technological innovations accompanied by islandwide reorganization of its river system.

For the most part, the infrastructural system, called the "hydrohub," controls water circulation and storage according to plan, allowing inhabitants and visitors to take it for granted. In such circumstances, the river, as a geological actor, becomes the background for more salient stages of human social interaction or remains completely backstage, so to speak, in the "unthought known" of the systems' unseen underground (Rubenstein and Russell 2010, 9). But when flash floods disrupt everyday life, the character of the rivers as "vibrant matter" brings attention to itself (Bennett 2009). The river, as a geological actor, comes to the front of the city stage, where the inhabitants, its audience, can't help but appreciate the power of its inconvenient presence.

In this chapter, I analyze frontstage/backstage shifts in the interactional dramas highlighting human and aquatic agency in two central Singaporean sites, Clarke Quay and Orchard Road. In both, the state—motivated by the efficient flow of capital and freshwater—transforms riverine structure, function, and meaning.[2] Based on fieldwork interviews, infrastructural site visits, and participant observation, I contrast two densely populated cityscapes that have been incorporated into the national reservoir system.[3] In the first site, the Singapore River continues to function in the cultural and economic life of the city, but having lost its place as the frontstage of everyday life and livelihood, the river is now relegated to serve backstage functions as a representation of cultural heritage and as water infrastructure. In the second site, an ancient river flows under what is now Orchard Road, Singapore's signature shopping district. A geological trace without contemporary cultural salience, the ghost river waits backstage in the deeper stratigraphic layers of its Holocene past, bursting forth on cue with intense, unpredictably localized rainstorms. Exceeding the capacity of the drainage system built into its former riverbed, the ghost river periodically takes the form of flash floods, disrupting commerce and transport.

As Basso (1996, 41) has argued, ethnographic study of landscapes points to "the symbolic attributes of human environments and the effects of environmental constructions on patterns of social action." The Anthropocene approach, however, requires rethinking the ethnographic landscape.[4] For it is not only humans who extract and project meaning upon the landscape; the earth's elemental forces are themselves actors that reconfigure the landscape. And indeed, rivers can disrupt our basic assumptions about terra—dry land—disruptions that structure our notions of the landscape and our modes of territorializing space (Kane forthcoming). The Anthropocene approach to aquatic flows also requires 3D geovisualization.

So in the case of Singapore, for example, the surface and ground waters run at different speeds but simultaneously through the streets, canals, and drains. In effect, they co-create the landscape with humans as they appear and disappear. Engineers systematically measure these flows through the hydrohub, adapting the islandwide water infrastructure to expected input and output rates calculated in reference to historical records. But they have no control over, or even a way of predicting, where and when sudden rainstorms will flash; climate change will only diminish the usefulness of historical data in their calculations (Whitington 2016). Surges of aquatic unpredictability can shock those with expectations of routine control of water resources even as they are lulled back into routine when the human-river time frames resynchronize.

This chapter engages with current research on meaning and action related to island surface topography by focusing on three-dimensional infrastructural arenas of urban river landscapes. Contrasting Clarke Quay and Orchard Road, I propose that collective human agency (here guided by a wealthy, pragmatic state) changes planetary history by moving and repurposing the material staging grounds—the riverine landscapes and infrastructures—of social interaction. The power to redesign the frontstage and backstage of human activity in the earth's crust exists in tension with the limits asserted by rivers. The tension around the dubious human power to predict and control rivers, I suggest, is a key feature of the Anthropocene.

WATER SECURITY IN SINGAPORE (GEOPOLITICS)

By the time Singapore became a sovereign republic in 1965, citizens had already elected the party that continues to dominate parliament today (the People's Action Party, PAP). The continuity of organizational decision making that this political formation allows contributes to the viability and micromanagement of infrastructural projects large and small. The government has made water a top national security priority, investing much of its great wealth in turning vulnerabilities (scarcity, pollution, flooding) into engines for innovation (Lee 2015; Lee and Ong 2015). The cleanup of the Singapore River in the 1970s—which entailed eviction of all the small boats called lighters (*tongkangs*)—was the first major attempt of its kind to improve water quality in Asia.[5] The lighters once carried cargo from the big ships in the port into the heart of Singapore's business district. Together with other traditional tradespeople who had lived and worked along the river, the people who worked the lighters played an important role in the island's development as a global center of maritime trade. Victims of urban-environmental renewal and the shift to containerized shipping, they had no choice but to leave (Dobbs 2002). Today, the Singapore River, empty of all but a few tourist and government boats, is integrated into the islandwide reservoir system while most of the people who once inhabited its waters live in public housing towers.

The Public Utilities Board (PUB) is the government agency responsible for integrating and managing the river as part of the island's water cycle. It has established Singapore as a global hub of research on water and development of water infrastructure (PUB 2012). The PUB has accomplished this engineering feat while "creating aesthetic waterways to enhance the urban environment" (Lim 1997), even as their central objective remains: to regulate the balance between a healthy water supply and flood control.

Producing Potable Water as Transnational Geopolitics

With two monsoons and no dry season, water scarcity in Singapore is not due to insufficient rain. Rather, there is not enough land to store the abundant rain. In 1961 and 1962, while still a British colony, Malaysia signed two agreements with Singapore assuring the continued transfer of water through three large pipelines across a 2 km causeway from the Malay Peninsula to the island. (Singapore buys raw river water, treats it, imports it, and also sells some treated water back to Malaysia.) When, in 1965, Singapore was expelled from the Malaysian Confederation, which it had briefly joined after freeing itself from the English, the separation agreement affirmed the two prior water transfer agreements. One ended in 2011; the other remains in effect until 2061. Although stable, the arrangement is a source of recurring political tension (Lee 2003). The neighboring Indonesian archipelago, which has supplied Singapore with labor and sand for land reclamation, is another potential supplier of raw water—and another source of political tension (Ong 2004). Anticipating future wrangling with its neighbors, Singapore is eager to centralize and diversify its water supply (Tortajada 2006). All of its own major rivers have been recruited in this effort. To keep the nation's freshwater separate from seawater, rivers have all been dammed and interlinked through a series of reservoirs. The many streams and creeks have, so far, evaded incorporation into the reservoir system, although they are directly in the sight lines of engineers who may eventually be able to tap all existing freshwater sources, even the smallest ones. Experimental desalination plants can now shift inputs from fresh to salt with changing conditions, thereby offsetting some of the higher energy costs associated with the desalination phase (PUB 2012). If the hydrohub is to become a global engineering model, the "enclave ecology" it creates requires critical scrutiny focusing on the consequences of banishing estuarial habitats (Kane 2017).

Techniques of Stabilization in Flood-Prone Topography

Prior to infrastructural transformation, forty basins drained the island's undulating topography of rounded hills. Local, brief, and intense rainstorms filled the narrow, short, steep streambeds of the basins, eroding the subdued spurs, carving gullies that carried sediment down the broken slopes to the valley bottoms where the floods spread in sheets across estuarine flats. Before all the post-1982 land reclamation projects, 20 percent of the island was below water at high tide (Gupta 1982). Urbanization has been accelerating these hydraulic processes, putting global city development and the accompanying creature comforts that engender citizen complacency (e.g., air-conditioning) at cross-purposes with water management. The resulting problems generate anxiety in the populace and in the government (Taylor 1934, cited in Gupta 1982).

Like urban rivers around the world, Singapore's rivers are hybrids of nature, culture, and engineering. Understanding the historically changing "techniques of stabilization" that sustain and reconfigure riverhood entails analysis of infrastructural practices and dispositions as they unfold differentially in space and time (Joyce 2010). Today's postcolonial authorities are not so unlike previous British colonial authorities who engineered "norms and forms" of the built environment to enhance the flow of economic activities through Singapore's port. These norms and forms function as "social technologies [and] as strategies of power to incorporate, categorize, discipline, control and reform" island inhabitants (King 1990, 9; see also Yeoh 1996). The mammoth islandwide drainage system, composed of over 803 km of concrete-lined conduits and earth drains (Lim 1997) is one such technique (or complex of techniques) of stabilization. Ubiquitous low-paid, non-Singaporean workers, subject also to the legal norms and forms of immigration, can be regularly observed carrying out the everyday micro-practices that keep the system free of debris.

As key actors, engineers in Singapore and elsewhere are beginning to shift away from the traditional "rational approach" to urban infrastructure that relies on the pure calculation of pathways (streams), sources (rain), and receptors (drains). They are shifting toward a more holistic (but no less rationalized) "risk management" approach that is more sensitive to the complexities, uncertainties, and struggles evoked by urban flood episodes. Balancing the probability that events will occur with the probable consequences should they occur, risk matrices delineate the biophysical, hydraulic, and weather signals used to guide decisions about whether and when to trigger intervention (e.g., raise floodgates) (White 2010; Task Force 2012). Citizens' reports and interpretation of localized flash flood events can support official decision-making processes based on risk matrices. PUB has created a website and phone apps through which citizens can participate; however, the extent to which their participation can meaningfully shift top-down, techno-managerial approaches is an open question.[6]

WHAT HAPPENED TO THE SINGAPORE RIVER? THE CHANGING MEANING OF RIVERHOOD

Intervention in the shape, function, and meaning of the Singapore River combines principles from engineering, architecture, aesthetics, and law. A continued focus of intense governmental intervention, the 3.0 km long Singapore River serves a central urban area of 3,707 acres and is a key element in the islandwide reservoir system. The century-old river walls that protected its banks were recently reconstructed to assure continued structural integrity. Refinished with granite, the walls also conserve the river's unique historical character, contributing to the ideological production of Singaporean identity as a landscape of global trade that has unfolded through colonial and postcolonial history. The bed was deepened to meet the drainage needs of the urban catchment area, and laws have been rigorously enforced to protect the drainage system (Lim 1997). Riverside quays were redeveloped to support contemporary tourist restaurants, bars, and related businesses in a uniquely functional blend of colonial British and tropical Southeast Asian styles that mimic the old to create a sense of distinction and continuity in the new. The rooftops, sidewalks, and decks on Clarke Quay, for example, coordinate patterns of crowd movement and rest and protect pedestrians from rain and sun. Architectural and infrastructural functions blend, such that a network of tubes channels freshwater directly into the river-reservoir system, preventing contamination by street pollutants (fig. 10.1).

Once river-dwelling traders were swept off the river's sanitized surface and tucked away in apartment towers and most commodities began to move through

FIGURE 10.1. The codesigned infrastructure-architecture of Clarke Quay captures rain and keeps tourists dry.

containers and truck docking nodes on the seacoast, the Singapore River no longer existed as the frontstage of everyday life. On Clarke Quay, as elsewhere, frontstage performance has shifted to land's edge. The river has become the backstage, or backdrop, to face-to-face transactions. As is usually the case when examining social life, everyday practices blur boundaries, including those between, above, or below the front- and backstage. In addition to their primary functions, infrastructures often perform an important though often unrecognized, secondary, unplanned function: infrastructures materially manifest socially *unassigned* spaces—even in the most controlled settings. Among the tightly organized city spaces, liminal spaces allow some escape. The space under a bridge, for example, may allow men to escape the gaze of employers, employees, and tourists. In the tightly regulated zone of Clarke Quay, they share a smoke and conversation offstage (fig. 10.2).

Such alternative possibilities in the cityscape go unmarked on maps. But the ethnographer can easily locate liminal spaces betwixt and between the crisscrossing infrastructural layers. In countries less disciplined than Singapore, graffiti writers and artists seek out the infrastructures that mediate more controlled zones because they provide material surfaces for illegal painting (Kane 2009). Although the Singaporean state restricts these outsider expressions, it nevertheless appropriates

FIGURE 10.2. Layered river infrastructure provides offstage social interactional space for people to escape more tightly surveilled front- and backstages of the Singapore River.

the creative power of graffiti style by hiring muralists to "activate" the pedestrian underpasses connecting different quays along the river with iconic, nationalist images of the river's rich history (fig. 10.3).[7] Given this opportunity, the graffiti artists bring the figures and events of Singaporean history back into this ghosted estuary. (For surely at high tide in rainstorms, before the hydrohub separated the river from the sea, the mix of fresh- and salt water would have once exceeded the current hardened banks and nourished the nonhuman creatures who once lived there.)

The material semiotics on the surface tell much about how the state engineers the river's cultural ecology. However, I argue that underground clues are imperative to understanding the techno-cultural formation of the twenty-first-century Anthropocene. The structures are so large and elaborate that they provide paleontologists with stratigraphic evidence marking human biosphere engineering (Williams et al. 2014). Beneath the architecture, infrastructure, and socio-techno-spatial organization of Singapore's surface, there is a well-lit, air-conditioned, parallel world of movement and habitation. Vast twenty-four-hour networks of consumption and transport host much of everyday life, providing relief from the exceedingly hot and rainy climate. It is underground, in the Clarke Quay Metro station, that the ethnographer comes across a material representation of the river's past: a replica of

FIGURE 10.3. Streams of images, pedestrians, and water intertwine along the underpasses linking touristic neighborhoods along the Singapore River. A nod to freedom of expression, the Singapore River One project appropriates the aesthetic power of graffiti art. This fragment interprets the "Lion City."

a series of four paintings by Chua Ek Kay, one of which depicts the displaced lighters haunting the surface of the once-busy river above (fig. 10.4). This river's history has been both usurped and preserved. It has vanished and then rematerialized as art: ink brush strokes on a screen; a pictorial landscape intruding like memory on the consciousness of a public taking escalators up and down the geological layers of time under the river. Thus, the river's longue durée is an element in the ideological production of a national past, a stratigraphic layer gesturing to the importance of what once was. Hydraulic engineering is also a way toward reimagining community (Anderson 1999). The remaking of place produces the nation, relying as always but differently on the Singapore River. The process of keeping the once socially active river trading zone alive as tourist backdrop, turning history into artistic material objects in various media, repeats in all the different touristic sections of the river.

The landscape encodes the multiple national objectives achieved through integrated storm water and freshwater management. Back on the surface of Clarke Quay, gazing downstream, the monumental Marina Bay Hotel and Casino looms over the skyline, an emblem of Singapore's nodal position in world financial flows. The triple-tower structure, linked at the top floor in the shape of a long boat, is a

FIGURE 10.4. Rendering the displaced lighters of the past for Metro riders: what was above, goes below; what were material transactions of everyday life becomes symbolic reminders. One of four "Reflections" by Chua Ek Kay, 2001, in the Art in Transit exhibit, Northeast Line of the MRT, Clarke Quay station.

key feature in the complex organized by the Marina Bay Barrage. The Marina Bay Barrage is a dam system with doors and pumps that opens and closes the hydraulic connection between the Singapore River, the reservoir system of which it is a part, and the sea (figs. 10.5a, 10.5b). In the development of the river-canal-reservoir system, water supply and flood control functions are enhanced for financial, recreational, aesthetic and environmental purposes (Lim 1997). The diversity of these urban spaces camouflages the systematized hydraulic connections among them. In fact, the Singapore River water flows down into the reservoir behind Marina Barrage, where it mixes with surface and underground flows from Orchard Road.

FLASH FLOODS ON ORCHARD ROAD'S CRUSTAL ACCUMULATIONS

Earlier in the geological epoch of the Holocene, the sea rose up and retreated from the islands of what would become Southeast Asia. Rain-fed rivers drained the valleys, and then the sea refilled the valleys with marine sediment (Gupta 1982, 138–39). Thousands of years later, in colonial times, a main thoroughfare was

FIGURE 10.5A. Looking downstream toward what was once the river's mouth, the Marina Bay Hotel and Casino represents and produces Singapore's moneymaking future.

FIGURE 10.5B. The Marine Bay Barrage regulates the island's floods and the freshwater catchment system.

FIGURE 10.6A. The frontstage designed for elite guests of a luxury hotel on a flood-prone bend in Orchard Road. Fragments of sculptures by Botero and Anthony Poon.

FIGURE 10.6B. A section of the Stamford Canal provides a backstage social interactional space for upscale hotel and mall workers to take a break.

established on top of the marine sediment of one such ancient streambed. In the nineteenth and early twentieth century, farmers carried fruits and vegetables to market along this thoroughfare, hence the name, Orchard Road. In the opposite direction, scavengers collected night-soil in buckets from residences and businesses, bringing it to the farms uphill—a practice that became a focus of contention between municipal colonial authorities and the city's Asian communities (Yeoh 2013). By 1980, Orchard Road was enveloped by the city center; humanity's crustal accumulations already blanketed 80 percent of the riverbed (Gupta 1982, 143). Subsequently, when the Orchard Road corridor was cleared for the Mass Rapid Transit (MRT) stations, a variety of religious temples, mosques, and chapels were demolished and relocated (Kong and Yeoh 2003). Like the lighters on Singapore River, the religious groups were given no choice but to make way for urban development. Today's Orchard Road hosts a stretch of mostly high-end, global palaces of consumption, residence, and business that together have become a symbol of national pride (Kong and Yeoh 2003). High art and architectural front spaces service elites, while infrastructural back spaces, like the Stamford Canal, provide a place for workers to take breaks (figs. 106.a, 10.6b).

Yet, despite wealth and futuristic engineering savvy, it is simply not realistic to expect flooding in Singapore to be totally eradicated. At any moment, freak storms coinciding with high tides cause havoc. The colonial-era Stamford Canal, reconstructed in 1978 and again in 1986, runs beneath and beside Orchard Road collecting and diverting surface waters. At the Marina Bay end, the open canal was closed off and now supports a spacious promenade. Its capacity, however, occasionally diminished by debris-clogged drains, can still be woefully insufficient (as in flash floods of 2010 and 2011). Localized flash floods from heavy tropical rainstorms may last less than an hour, yet still cause vehicles to float away from their parking spaces (Lim 1997). The PUB encourages owners of larger buildings to invest in computer-monitored flood walls that protect key entrance space as well as in internal water storage tanks that temporarily hold excess back from the public drainage system, giving it a chance to clear.[8] Many people who work in the more flood-prone parts of Orchard Road think that the state should provide sufficient, effective infrastructure so that they are not victimized by flash floods and so that the futuristic image of Singapore as a twenty-first-century mecca for the rich and aspiring is not muddled by wading shoppers and ruined (not always insured) merchandise, furnishings, and equipment.

As an entity without cultural salience (except among scientists, including social scientists), the river that shaped the topography of Orchard Road has disappeared, ghostlike, into the ancient past. When intense rains fall into the valley sculpted by this prehistoric river, the rain and topography summon this ghost. Flash floods can assume the force and form of the lost river and challenge the existential premise that the geological deep is fixed in our past.

CONCLUSION

Can there be a good Anthropocene?

Jai Syvitski, "The Anthropocene—from Concept, to Geological Epoch, to 21st-Century Science and Public Discourse"

Vast investments and hugely creative and destructive technology can drive back the reckoning, but cheap nature really is over.

Donna Haraway, "Anthropocene, Capitalocene, Plantationocene, Chthulucene"

Building our environments into the crust of the earth, humans revise the planetary surface, shifting the stage of social interaction in dynamic relationships with aquatic flows (Kane 2012). Singapore's futuristic engineering of island water, its nearly complete transformation of hydrology into hydraulic engineering, is a tiny piece of the larger puzzle of the global transformation of the planet's river systems (Meybeck and Vörösmarty 2005). The tightly governed, wealthy island city-state provides a particularly illuminating case of the never-ending tension between the quest for control (here managed with unmatched technical efficiency) and the chaotic possibilities inherent in technologized human-river relationships (here pinpointed with frustrating exactitude). There are an uncountable number of parallels variably enacting this infrastructural tension in the world's cities. In the nineteenth and early twentieth century, when most cities engineered their way into modernity, they did so in part by covering rivers, often selected for the degree of sewage effluent polluting their courses. Rivers were rerouted under streets, enhanced with electric and water distribution networks, and lined with architecture, and in the best of places, enhanced with art. Many cities unintentionally built central streets on paths sculpted by ghost rivers. This—and all that came before this—is still happening. We can track the dynamics of riverine appearance and disappearance. Researching the frontstage/backstage shifts across geological and historical time can reflexively inform science, storytelling, art, and policy. As the shifting relationships between rivers and cities unfold in the places studied and in the scholarship itself, the dynamics may inspire new ways of imagining the future. (Mary Miss's installation art comes to mind for the ingenious ways it brings the taken-for-granted White River literally into personal reflections of city inhabitants.)[9] Filtered through the different modes of knowledge and media production, the focus on front- and backstage urban transformations and riverine ghosts are simple analytic tools that can align inquiry, representation, and action in the transdisciplinary space of the Anthropocene.

ACKNOWLEDGMENTS

Thanks to Jason Kelly for organizing and inviting me to the Rivers of the Anthropocene conference, at IUPUI Arts and Humanities Institute, Indianapolis, in January 2014. Thanks to participants for discussion, especially Jan Zalasiewicz for introducing me to Anthropocene landscapes as scientific ghost stories. In Singapore, thanks to my interlocutors and hosts at the National University of Singapore, especially David Higgitt in the Department of Geography, who steered me toward Orchard Road. Special thanks to Harry Seah and Chow Qin Wei in the Technology Department of PUB for making the agency's engineers and infrastructure sites accessible. Thanks to Edith Lea Hernandez and Andres Eskjær Jensen for sharing their lovely home. Thanks also to participants of the Infrastructural Worlds workshop at Duke University in March 2014, especially the Waterways and Coastlines break-out group, Majed Akhter, Ashley Carse, Joshua Lewis, Ben Mendelsohn and Chitra Venkataramani, for discussion of an earlier version of this chapter. Thanks to Nur Amali Ibrahim, my International Studies colleague, for insightful discussion of Singaporean culture and politics. And as always, C. Jason Dotson, for project support and videography.

NOTES

1. Goffman's (1959) theory of frontstage and backstage concerns the construction of self and the performance of identity. Individuals manage information about themselves in a variety of ways, one of which is moving into and out of different settings, e.g., taking off makeup backstage after performing onstage. I extend these terms to analyze the dynamic and layered shifts across spatiotemporal scales. I upend Goffman's assumption that "a setting tends to stay put" (24) to consider how states (and other entities) reassemble entire settings (and their infrastructures), switching up activities in the frontstage and backstages as they globalize local capabilities. Graham (2010, 18) also finds Goffman's frontstage/backstage metaphor useful when analyzing disruptions caused by urban infrastructure.

2. As actors in networks, meshworks, and assemblages, humans can calculate and intend but not control. Material forms lacking intention, like rivers and rock, can shape events in linear ways that conform to engineering models or in nonlinear ways that appear chaotically. The predictability of their agency, or action in the world, varies with the situation. For more on this approach, drawn from the social study of science and technology, see Mitchell 2002 and Latour 2005; as applied to seismic science and communications infrastructure in a tsunami, see Kane, Medina, and Michler 2015.

3. The fieldwork project in Singapore (conducted in May 2013) is one in a series of studies of water management in the context of environmental change. The chapter draws on data from sixteen interviews with scholars and engineers and twelve visits to key infrastructure sites. For further detail on ethnography of infrastructure methods, see Kane 2017.

4. Thanks to Ashley Carse for this insight.

5. Interview with David Higgitt, Department of Geography, National University of Singapore, May 6.

6. http://www.pub.gov.sg/managingflashfloods/Pages/default.aspx and http://www.stomp.com.sg/. Both accessed 3/8/14.

7. See www.singapore-river.com. Accessed 3/8/14.

8. Interview with building supervisor, Orchard Road, May 13.

9. See http://www.imamuseum.org/visit/100acres/artworks-projects/flow, http://www.marymiss.com/index.html (accessed 7/6/16); and Miss and Carter, this vol.

11

Decoding the River

*Artists and Scientists Reveal the Water System of the
White River*

Mary Miss and Tim Carter

It has been become apparent over the past several decades that we are facing issues of increasing urgency in relationship to the environment in general and our water systems in particular. These natural life-support systems are often taken for granted and in many cases have become invisible to urban and suburban dwellers. The adoption of the term "Anthropocene" in the urban context recognizes the fundamental role humans have in creating, manipulating, and shaping the water systems and the environment during the very recent past. Relative to nonurban systems, the ecology of the urban water system is highly modified, and many measures of water quality are affected negatively. Scientists have become frustrated that their ongoing research into the effects of this degradation and the impact it will have in the future goes unnoted by the broader public. In the past few years artists and scientists have begun collaborating to create projects that will begin a process of engaging the general public with these pressing issues. The goal is to create awareness that leads to action and the development of more sustainable communities. Following is a description of the process of creating that engagement in two consecutive initiatives focusing on the White River and its tributaries. A replicable model or approach is envisioned that will promote inquiry, encourage participation, and help citizens become part of the "green infrastructure" of their cities.

The White River presents itself day to day as a bucolic stream as it winds its way through the city of Indianapolis. Part of the Ohio and Mississippi River systems, but too shallow to be navigable, it is unindustrialized for most of its length. Where it reaches the center of the city, factories have traditionally used it as a water source and for waste removal. The storm water infrastructure of the city also relates importantly to the health of the White River. The lack of significant

SUSTAINABILITY MADE TANGIBLE THROUGH THE ARTS

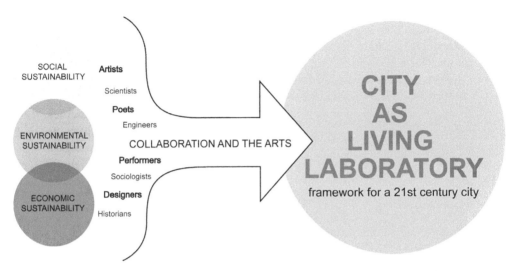

FIGURE 11.1. City as Living Laboratory (CALL) framework diagram. Image courtesy of MM/CaLL Studio.

topographic features makes drainage a challenge. Flooding and standing water on the streets after rainfall is common. How could this river system, which supplies up to 70 percent of the drinking water for the city, become more visible to its citizens? In the approach to this question, it is essential to create situations and creative encounters through which residents and the general public become aware of the infrastructure that moves the water through their city, visualize the habitat corridors that streams and rivers provide within core urban neighborhoods, and understand the role water has in the economic vitality of the city.

The City as Living Laboratory: Sustainability Made Tangible through the Arts (CALL), an initiative developed by Mary Miss and Marda Kirn, is a framework intended to be used to make issues of social, economic, and environmental sustainability compelling to the public (fig. 11.1). It envisions the city as a laboratory in which collaborations among artists, scientists, planners, and communities can make a city's pressing issues apparent to its citizens through projects and events. This method of translating the city is beginning to be studied and evaluated for its effectiveness, and as the emergence of these collaborative practices continue to be executed, each case study can provide new lessons to help inform and shape future outcomes.

FLOW (Can you See the River?) is a project of the CALL framework, commissioned by the Indianapolis Museum of Art. FLOW begins with the assumption that "all property is riverfront property—the river starts at your front door." The intention is to engage viscerally the citizens of Indianapolis and make them aware of the multiple ways the river and its watershed support their lives.

Working with scientists from the U.S. Geological Survey, Butler University, and Indiana University–Purdue University, Indianapolis, a series of installations were implemented in 2011 from the grounds of the Indianapolis Museum of Art to the White River State Park, six miles to the south in downtown Indianapolis. The installations were intended to engage people's interest in the complexity of this familiar feature of the landscape that they often take for granted.

As people moved along this six-mile stretch of bike paths and parklands, they encountered a series of stopping points. At each point a new aspect of the river—its infrastructure, history, or ecology—was revealed. Visitors could choose to engage in passing or choose to get in-depth information from a dial-up number, website, or app developed for the project. These stopping places were modest in nature, like acupuncture points that accessed different aspects of the circulatory system that is the White River. In addition, FLOW used collaborative community network programming to disseminate the project's messages.

GOALS

There were very specific project goals for FLOW. We hoped to expand public awareness of the White River watershed to let people know what it is, how it functions, and what it means to Indianapolis environmentally, economically, and socially. We wanted to help citizens begin to understand their actions—at home, at work, at school, or at play—in direct relationship to the river upstream or downstream. Finally, we also hoped to inspire new learning and collaborations among individuals, institutions, and agencies that could lay the foundation for future activities.

ELEMENTS

A locator/marker was developed to direct attention to specific aspects of the river. It took the form of a circular, stainless-steel mirror attached to a pole that was positioned to reflect a particular point of focus—a levee, a storm water drain, or a wetland. A red sphere/marker, like an enlarged pin on a map, was placed on the location of interest. A red mark was made on the surface of the mirror (fig. 11.2) As the viewer aligned the red marks, his or her point of focus vibrated back and forth between the surface mark and the reflected sphere/marker. Text etched on the mirror surface identified the point of focus. These mirrors appeared singly or in clusters and in a variety of sizes. Looking at their own reflections in the mirrors, viewers could see themselves in relation to the river (fig. 11.3)

On the surface of the stainless-steel disc, a dial-up number was given from which the viewer could hear a brief description of the point of focus. A website address was also given where it was possible to read about the topic in more detail.

FIGURE 11.2. FLOW (Can You See the River?). Diagram illustrating mirror's reflection of red markers designating points of focus in the landscape—the river, a tree, the wetlands, and so on. Image courtesy of MM/CaLL Studio.

A walkable map, approximately 40 feet square, was located in the entry pavilion of the Indianapolis Museum of Art. With playful, oversized red balls scattered around the surface, viewers were invited to engage in locating their own home, school, or business and to see their relation to the river and the tag line "All Property is Riverfront Property" (fig. 11.4).

Butler University secured a grant from the National Oceanic and Atmospheric Administration (NOAA) and worked closely with Mary Miss Studio to develop a web application, www.trackaraindrop.org, that made it possible to track the movement of a drop of water from any place in the city to the river. It allowed the user to see how the water flowed in rain events of different intensities, how the water got to the river (a pipe, open channel, or stream), and what the pollutants were in that particular area. The app also made it possible to find the difference between weather and climate and what individuals could do to help clean up the river.

TOPICS

The types of topics addressed in the project were varied and intended to engage a variety of interests. For example: What is a watershed, and what is the White River's relationship to it? How does water circulate, and is there any "new" water?

FIGURE 11.3. FLOW (Can You See the River?). Image courtesy of MM/CALL Studio.

What are the local stories about the river? How does the river shape the land and the land shape the river? What are floods, and how do they happen? What is the ecology of the river, and what affects it? What is the history of the river?

A series of events were organized with community partners for a White River Festival. We imagined this community as the "human ecological infrastructure" of the city, which was a term we adopted to refer to established organizations that were already working to safeguard and improve the environment of greater Indianapolis such as government agencies, scientific institutions, and cultural organizations. These groups organized the White River Festival to align with the opening of FLOW. Activities included discussion panels, dance performances, tours, exhibits, and talks.

OUTCOMES

The FLOW prototype was the first, full implementation of the CALL framework and was a remarkable opportunity to test the use of a variety of strategies. One informal lesson that we took away from this project was that multiple means of access are essential to engage the most diverse group of visitors—dial-up, website, apps, and events. We felt that Raindrop, for example, could potentially engage more

FIGURE 11.4. FLOW (Can You See the River?). Walkable map of the city of Indianapolis that enables one to locate one's own home in relation to the river, illustrating that "all property is riverfront property." Image courtesy of MM/CALL Studio.

people via additional programming and dissemination through formal school curricula. Through our experience here and on a subsequent project (Broadway: 1000 Steps, www.broadway1000steps.com), we began to understand that events that happen repeatedly over a period of time are the most effective way of drawing more people in at a deeper level.

We evaluated the project using quantitative and qualitative methods. Quantitative results demonstrated that both FLOW and Raindrop had varying

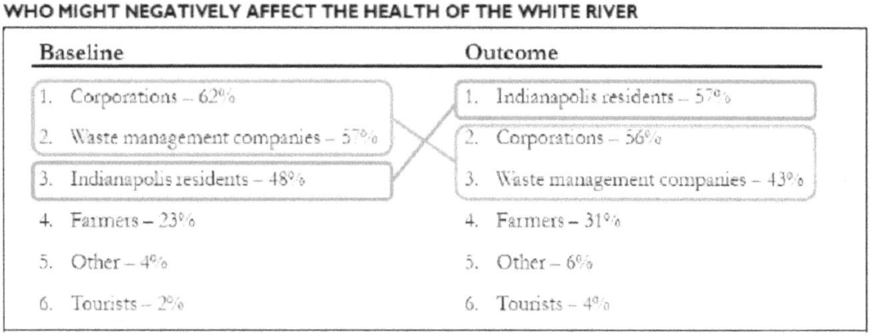

WHO MIGHT NEGATIVELY AFFECT THE HEALTH OF THE WHITE RIVER

Baseline	Outcome
1. Corporations – 62%	1. Indianapolis residents – 57%
2. Waste management companies – 57%	2. Corporations – 56%
3. Indianapolis residents – 48%	3. Waste management companies – 43%
4. Farmers – 23%	4. Farmers – 31%
5. Other – 4%	5. Other – 6%
6. Tourists – 2%	6. Tourists – 4%

FIGURE 11.5. Sample FLOW evaluation results.

levels of effectiveness in raising awareness about water-related content about the White River (fig. 11.5). The attitudinal data that we collected which focused on awareness of the White River demonstrated little change before and after experiencing FLOW—with none of the changes to the metrics having a statistically significant difference. The baseline respondents reported that the White River was of high importance to the city of Indianapolis, and this "ceiling effect" made significant changes between baseline and outcome cohorts difficult to observe.

Qualitative interviews told a more nuanced story regarding awareness of the White River. After engaging with the project, interviewees reported that they learned about the river, and this translated into a meaningful personal experience. For example, one respondent said, "Before participating in the project, I had no idea that there was a hundred-year flood, so I definitely was educated about the history of Indianapolis.... So seeing the red markers and the red balls and the mirrors and everything was a harsh personal context for me." Another reported, "It [FLOW] made me think about really where the role of the River is in our community, and again, how hidden it is in places. It's caused me to lament the fact that when I cross the bridges that I cross day to day, I lament the fact that I can't see the River" (RK&A 2012).

Raindrop was also evaluated using interviews. It was most clearly defined as an educational resource, and respondents described its value as raising personal awareness of the water systems in Indianapolis. This included statements such as "It prompted me to be more aware of what's going on around me. It makes me realize how much environmental issues are going on here and how much there needs to be a change in our behaviors"; and "I think [Raindrop] raises everybody's awareness that everything flows into [the White River] and everybody needs to pay attention to what they're doing with their water." The nature and frequency of responses indicated a successful connection between the intended outcome of the app and actual user experiences.

Perhaps the most compelling outcome of this initiative was its ongoing effect within the community. At the beginning, we spoke about our intentions for this project; it was intended to be modest in form, not dependent on spectacle. We hoped that it would be a catalyst within the city, to help start other initiatives that elevated the value of the waterways of Indianapolis in the eyes of the public.

We have built on our experience with FLOW and continue the process of connecting the citizens of Indianapolis to the White River. In 2013, another collaborative CALL project—Streamlines—was funded through the National Science Foundation's Advancing Informal STEM Learning (AISL) program. The lead project team included representatives from Butler University, Indiana University–Purdue University Indianapolis, and New Knowledge Organization Ltd.; and the overall partnership included representatives from Indianapolis-based organizations such as Reconnecting to Our Waterways (ROW), the Indianapolis Museum of Art, and the DaVinci Pursuit. This project, launched in the fall of 2015, focuses on tributaries to the White River in five neighborhoods that were originally identified by ROW and uses four art forms (visual, music, poetry, and dance) combined with relevant, water-based science content to create informal learning sites in the city. A community facilitator from the Streamlines team worked with every neighborhood to identify general areas where the projects could be implemented. This dialogue with the neighborhoods is an ongoing part of the process.

Miss's role has been to develop a conceptual framework for this project that creates sites that will be activated by multiple artists and with community partners through ongoing events. Each site was chosen because of its specific characteristics through a dialogue between the artists and a group of scientists. Interpretive themes range from focusing on habitat corridors, water infrastructure, atmosphere, and land use to water as a resource and change over time. These themes in turn are associated with "keywords" such as precipitation, infrastructure, temperature, contamination, restoration. Some of these "keywords" are shared between sites, while others are specific to single locations (figs. 11.6, 11.7). The most salient topics at each site are noted, and visitors explore them through a series of on-site interventions, virtual devices, and programs. A series of prompts encourages each person to seek out specific aspects of each site through a kind of game-based wandering—what might be called a *dérive* or "ludogeography" after the work of Guy Debord and the artists Nikki Pugh, Ana Benlloch, and Stuart Tait, respectively (Dubord 1955; Benlloch, Pugh, and Tait 2008). An onsite map shows all five locations and their topics as well as the keywords associated with them. Visitors are encouraged to construct their own tours of sites according to their interests. For instance, if "habitat" is their choice, there may be three out of five sites where that topic is the focus.

As part of the project's conceptual development, we have created iconography that takes the form of a splayed star, which is referenced in the different layers of the project. Uses of this iconography include graphic identity, suggesting the relationship of topics at the five locations, and forming structures on the sites to prompt visitors to move out and explore the surroundings.

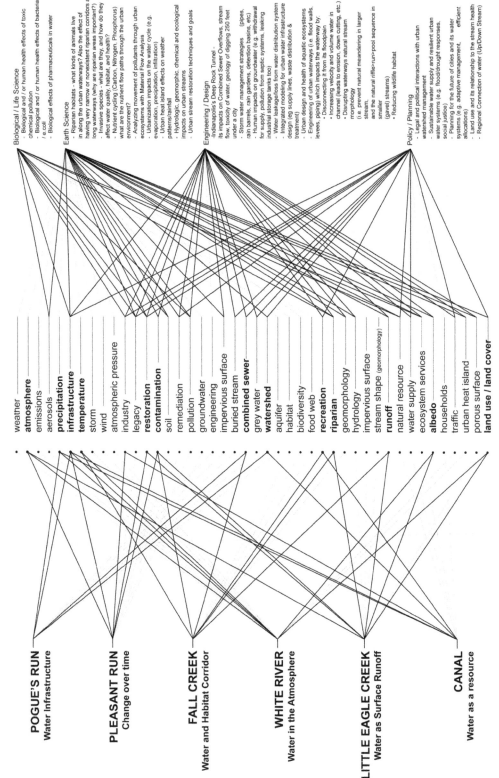

FIGURE 11.6. STREAM/LINES (I/CALL). Diagram illustrating keywords for each of the six tributary sites off the White River and their interrelationship. Image courtesy of MM/CALL Studio.

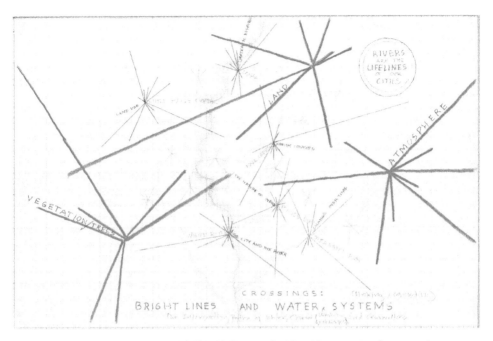

FIGURE 11.7. STREAM/LINES (I/CALL). Drawing by Mary Miss mapping the connections between the water system and the city of Indianapolis. Image courtesy of MM/CALL Studio.

The sites are activated in a number of ways. Interventions by Miss are encountered in the form of markers and mirrors, areas of plantings for cleaning storm water, or places to sit and reflect on the issues that affect our waterways. Music specifically commissioned for each site is accessible virtually. Texts composed by Indiana poets appear on the site, and site-specific dance performances that involve the community are being held over a two-year period.

The intention is to give communities adjacent to each of the sites a better sense of how their homes, streets, and businesses are connected to the river system and how important it is in supporting their daily lives. The goal is to arouse curiosity and a desire to visit all five locations. These sites, in combination, reveal multiple aspects of the city's water system. By dispersing sites around the city, we aim to initiate new levels of water awareness throughout Indianapolis.

CONCLUSION

The Anthropocene context is one in which hybrid ecosystems, like the urban water system, are complex and largely hidden by design. These systems make awareness and care a major challenge but very fertile territory for collaboration among artists, scientists, communities, and policy makers. Our work thus far has helped to

Principles for connecting knowledge, perspectives, artistic interventions with actions to promote sustainable development (working draft ver1.0 for Streamlines)

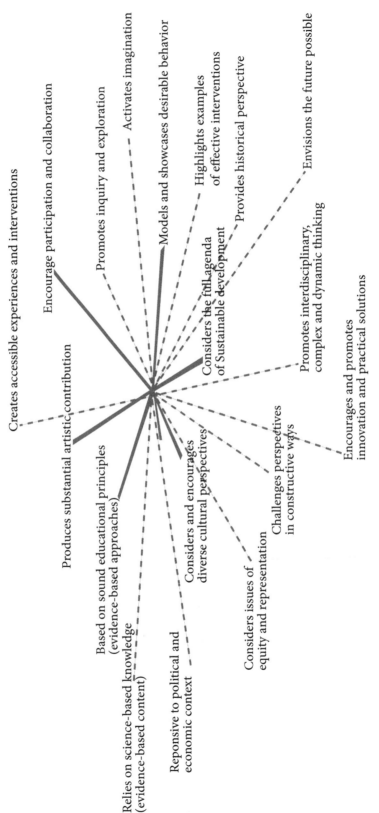

FIGURE 11.8. Principles for connecting knowledge, perspectives, and artistic interventions with actions to promote sustainable development (working draft v1.o for Streamlines).

FIGURE 11.9. STREAM/LINES (I/CALL). Installation at Butler University.

expand on CALL's premise that the arts can be an effective way to communicate science and reveal to individuals the ways in which they are embedded in their "environment." FLOW was a demonstration of how this could be done with modest installations along a major waterway. "Streamlines" expands this initial work into a citywide initiative (figs. 11.8, 11.9). A major goal for CALL is to develop a replicable framework that can be used by other cities to address the multiple challenges we face, particularly in our rapidly expanding urban environments. More types of projects are necessary for us to understand the ways these art-science collaborations can be most effective. Scientific research and government regulations alone will not be enough to help us deal with the challenges we face in the Anthropocene epoch. Finding ways to encourage citizens in all types of communities to engage in sustainable development in the face of climate change is essential to maintaining resiliency, health, and equity in our cities. Projects like those described above are intended to activate the imagination, encourage participation, and make it possible to envision a sustainable future.

12

What Is a River? The Chicago River as Hyperobject

Matt Edgeworth (narrative) and Jeff Benjamin (photos)

What is a river? Dictionaries define rivers as large natural flows of water, crossing or surrounded by land, flowing into an ocean or lake. The common assumption that rivers are natural entities—part of pristine natural cycles and processes—is deeply engrained. But contemporary rivers, as this book and other studies have shown, are far from being wholly natural. On the contrary, they have typically been subject to extensive sculpting and shaping by human beings. The question therefore arises as to whether rivers should be regarded as artificial instead. But that would be equally misleading, for biological and geomorphological processes are still at work even in the most controlled rivers. To insist on seeing rivers as either natural or artificial would be to reproduce entrenched dualistic frames of thought no longer applicable to understanding the hybrid entities of the Anthropocene.

Let us say instead that rivers are complex entanglements of artificial and natural forces—hybrid forms that are neither natural nor cultural, neither human nor nonhuman, neither social nor material, but confluences or mixtures of all these. They can accurately be characterized as "organic machines" (White 1996) or "cyborg-like environments composed of an interconnected and interdependent web of natural and artificial parts" (Scarpino 1997, 5). It might even be argued that human-influenced changes to rivers globally are so great that they helped bring about a new evolutionary stage of rivers in geological terms (Williams et al. 2014).

The study of rivers on a global scale has been facilitated by development of computers and GIS software such as Google Earth. Many researchers in various disciplines now encounter rivers principally via computer screens. Although there are considerable advantages afforded by computer technology, however, a problem is the lack of physical engagement with rivers entailed by studying them remotely. For all that is gained through virtual observation and analysis of riverine evidence on multiple scales, something of the material reality of rivers is lost. The force and vibrancy of river forces need to be experienced directly on an embodied level and scale of experience too, which is why a phenomenological approach is adopted in this chapter. "Phenomenology" here simply refers to the study of phenomena *as directly experienced*. A canoe on the river, rather than a computer console, mediates river encounters.

Imagine an explorer setting off by canoe to explore the natural wonders of the Amazon River. Such an idea (reenacted occasionally but persistently in exploration-themed television programming and geography magazines) appeals to our conceptions of nature as being all around us, as though culture consists of small islands enclosed on all sides by the ocean of nature. But now turn that around and inside-out. The concept of the Anthropocene entails the growing awareness that human culture and technology infiltrates so-called natural systems to a much greater extent than was ever imagined before. We now know, for example, that neither the rain forest of Amazonia nor the river system that supports it are quite as pristine or natural as once assumed (Raffles 2002; Schmidt et al 2014): humans have been thoroughly embedded in forest and river ecosystems for thousands of years. It is still possible to set off on riverine voyages of discovery by canoe, adopting the same spirit of curiosity and adventure that might be taken up if one imagined the river to be a pristine environment. But the wonders to be encountered are the cyborgs and hybrid entities mentioned earlier. Any canoe trip along a river is necessarily a journey through a complex and multifaceted reality, irreducible to parables about nature.

This chapter recounts such a journey, albeit a short one. It tells the story of an encounter with one particular river of the Anthropocene—the Chicago River. Ostensibly a minor river system within a relatively small watershed, it is described here as part of something much larger and more difficult to grasp: a hyperobject. Hyperobjects, to make of a concept recently developed by Morton (2013), are understood to be "massively distributed in time and space relative to humans," so huge they can never be apprehended in anything like their totality (1). They are part artifacts in the sense that humans have played a role in bringing them into being. In a useful appraisal of the potential relevance of the concept to archaeology, Hudson (2014) refers to them as "dark artefacts." Global warming might be held up as the classic example of a hyperobject. Inadvertently influenced by the industrial activity of human populations in the past, it could conceivably be intentionally shaped in the future. But that does not mean it is or ever could be entirely under human control. It can act independently and unpredictably. It can develop along trajectories that are unintended and unanticipated, and can phase in and out of human experience in unexpected ways.

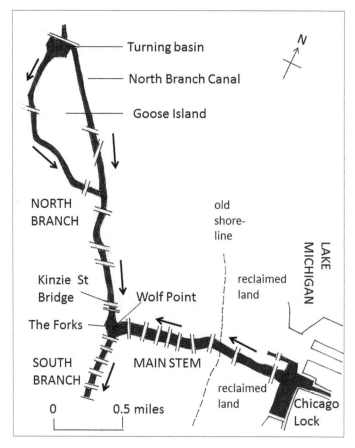

FIGURE 12.1. Map of the North Branch, South Branch, and Main Stem of the Chicago River, showing places mentioned in text. Arrows indicate direction of flow.

BACKGROUND

It was in a spirit of adventure that two of us set out to explore a short stretch of the Chicago River by canoe. It is common knowledge that the direction of the river's flow was artificially reversed in the 1890s (e.g., Solzman 1998). To anyone interested in archaeology of flow (Edgeworth 2011) this makes the Chicago River worth investigating further. The occasion to do so presented itself in May 2013, while visiting the University of Chicago for a Theoretical Archaeology Group (TAG) conference session titled "Archaeology of the Anthropocene." Archaeologist and artist, Jeff Benjamin, traveled from Michigan with his canoe on top of his car. It is a handmade wooden canoe, not dissimilar to the craft that would have been seen on the river before the city of Chicago was built. At 16 feet long, it can accommodate two people comfortably. The day after the conference session, with issues of the Anthropocene still fresh in our minds, we took the canoe to the northern suburbs of the city, where we could gain access to the river.

RIVER EXPLORATION

We put the canoe on the North Branch of the river at the north end of Goose Island, where there is a turning basin once used for industrial barges. It is the broadest stretch of the Chicago River.

Pushing off from the bank puts you in touch with currents acting on the boat. The flow of the river orients you, and you start to orient yourself in relation to upstream and downstream. Using paddles to propel and steer the canoe places your own human agency—the muscular movements of the body—into an active engagement with river forces. Through the medium of the boat and the paddle you come into contact with the vibrant, flowing materiality of the river.

Heading downstream towards the city, there is a choice as to which way to go. We take the canal route down the east side of the Goose Island. This part of the river is fairly shallow and there are no other boats on this stretch today. The water is smooth. The city lies before us. It is a gentle introduction to the Chicago Area Waterway System.

Goose Island sounds like a "natural" place, but it was formed by the cutting of the North Branch canal in the 1850s, bypassing a bend in the river for barge traffic. The island thus created became a huge industrial complex known as "Little Hell," lit up all night with blast furnaces and rolling mills (Solzman 1998). River frontage on two sides facilitated movement of goods and materials by barge. It was not just the North Branch canal, but the whole course of the river that was canalized—straightened, deepened, widened, dredged, and embanked—to allow passage of boats. Today the North Branch canal has partly silted up, "renaturalizing" itself, even if it was not natural to start with. Geese and other waterbirds, flying low and skimming the surface of water, use it as a kind of natural corridor through the city.

Rejoining the main course of the North Branch we go under an increasing number of bridges as we get closer to the city center. Chicago is famous for its movable bridges, with bridge towers for housing lifting or pivoting mechanisms. Of particular interest is the Kinzie Street Bridge. On the south (downstream) side are two fender piles, each consisting of multiple wooden stakes bound together, driven vertically into the riverbed.

It was here in 1992 that an unusual event occurred. An eddy of water and debris several meters across was observed in the river next to these piles, like water going down a very large plughole. At the same time, rising water was noticed in the basements of nearby buildings. It became clear that the river was emptying into a largely forgotten and disused system of freight tunnels, 60 miles in total length, that connected to basements in the city center. A state of emergency was called, and much of the Loop area had to be evacuated (Wilkerson 1992).

FIGURE 12.2. Heading downstream on the North Branch canal, toward the city center. Photograph by Jeffrey Benjamin.

It later transpired that workers putting in the fender piles the previous year, to strengthen and protect the bridge structure, had inadvertently damaged a section of old freight tunnel 20 feet below the riverbed, pushing displaced clay into the old tunnel wall. A small leak developed, which gradually got worse over the course of several months. Water seeping through the damaged tunnel wall increased until a small hole was created. The flow of water into the tunnel eroded the sides of the hole further until it was several meters across. Hundreds of millions of gallons of water went through the hole into the tunnel system, and from there it started filling up city basements, causing damage that cost over a billion dollars to fix.

It is worth noting that unseen and unsuspected events far below the surface (the flooding of the tunnels) and the existence of subterranean spaces and structures (the disused freight tunnel system) can be indicated by flow patterns on the surface of the river (the eddy of water and debris). The relevance of this will become clear in due course.

Arriving at the confluence of the North Branch, the South Branch, and the Main Stem, we find ourselves in yet another broad turning basin for ships and barges. The confluence has been widened far beyond its original dimensions. A huge iron barge is moored along the western side to our right.

FIGURE 12.3. Wolf Point: (a) view up the Main Stem (b) beached materials. Photographs by Jeffrey Benjamin.

The confluence is at the center of the "Y" that appears so often in civic symbols—the so-called municipal device—with the stem and two arms of the letter representing the three river branches. Sometimes known as the Forks, this part of the river is in many respects the symbolic heart of the city. Two hundred years ago it was surrounded by creeks and swamps, with a few log cabins. The first bridge over the river was here. Several decades later it was bordered on all sides by lumberyards and stockyards. Now the lumber and cattle have gone. Riverside plots afford prime land for property developments.

We look around for a place to pull into the bank and take in views of the skyscraper city. Finding a good spot is difficult. Along the bank on the left-hand side are underwater forests of thin vertical timber piles with sharp points sticking up just below the surface of the water; we push away from these with our paddles. Eventually we find a way through to the timber frontage of an old wharf—a relic of the time when the riverbank here was the center of the logging industry, and ships stacked with timber from now-vanished forests used to dock here. We moor the canoe, sit on the rocks (some of them are actually lumps of concrete), and eat our lunch. We have arrived at Wolf Point.

Wolf Point is a good place from which to consider the incredible transformations the river has gone through. Originally the Chicago River flowed *into* Lake Michigan, the source from which city drinking water was taken. But in the late nineteenth century the pollution from sewage and industry and meat production got so bad that lake water became dangerous to drink. The radical solution was to *reverse the flow* of the river—so that it would flow *away* from the lake. This was achieved partly by the building of the Chicago Ship and Sanitary Canal from the South Branch River into the Des Plaines River to the west. The canal was made progressively deeper the farther from the city it went, drawing the waters of the river into it (Solzman 1998). That meant that Chicago's industrial and sewage effluent flowed into the Illinois and Mississippi Rivers and ultimately all the way to the Gulf of Mexico. It still does. In linking with the larger Mississippi watershed, the Chicago River became part of a greater reality, almost continent-wide.

This has recently given rise to the problem of invasive species threatening to cross over watershed boundaries. The advance upriver of bighead and silver carp is an example. These voracious feeders were originally bred for their great size and rapid growth in Asia, where they were farmed for food over thousands of years. They were introduced into the Deep South to help clean up sewage ponds and commercial fish farm lakes, but some escaped into the river. Now huge shoals are heading up the Mississippi and Illinois Rivers, their numbers multiplying rapidly, threatening to break through electric barriers into the Chicago River and from there into Lake Michigan (Theriot and Tzoumis 2007). It used to be that the pollution in the river was so bad that it formed a toxic barrier through which no living thing could pass. But though still teeming with fecal bacteria, the river is cleaner now, and the formerly impenetrable barrier no longer holds. If the carp get through, they are predicted to transform the fragile ecology of the Great Lakes. There is now talk of reestablishing a more substantial physical barrier, effectively separating river watersheds that were artificially joined over a hundred years ago (Hinterthuer 2012). This would entail, among other things, engineering the re-reversal of flow in the Chicago River.

The skyline viewed from Wolf Point is spectacular, but as archaeologists our eyes are also inexorably drawn downward to the ground beneath our feet, as our attention alights on the mundane mixture of materials there.

It is a beach, not of sand and shells, but of artificial and natural materials, some washed up by the river in flood, held in place and stopped from slipping back into the water by the row of half-broken vertical piles. Many different kinds of humanly modified materials are to be seen here among the flotsam and jetsam—plastic bottles and floats, lengths of nylon rope, leather soles, planks and stakes of wood, strips of textile, styrofoam cups. It is a typical assemblage of Anthropocene objects but sorted by the river, thus weighted in favor of things that float and have been carried by the current.

Setting off again from Wolf Point we head up the Main Stem of the river toward Lake Michigan. On either side are soaring cliffs of concrete, metal, and glass, obscured from view only when we pass under the great underbellies of movable bridges.

FIGURE 12.4. Skyscraper canyon: heading upstream on the Main Stem. Photograph by Jeffrey Benjamin.

Chicago has always been a river city. City and river are gridded into each other, are parts of the same larger entity. Where is the floodplain of the river? It is integrated into urban architecture and infrastructure. If the river overflows its rusty metal banks, it fills the basements of buildings downtown. Where are the tributaries? The streams that once fed into the river have long since been culverted and incorporated into the system of sewers. Where is the catchment basin? The catchment is a concrete one. When it rains heavily, the impervious vertical and horizontal surfaces of the city—rooftops, windows, streets, curbs, parking lots, gutters—collect and channel storm water directly into the sewage system, instead of absorbing and gradually releasing it as the old wetlands and marshes did.

Now we are heading upstream against the direction of flow (famously reversed) through the skyscraper canyon that is the Main Stem. There are many more vessels on the water here in the city center—speedboats, barges, boats carrying tourists on river architecture tours. The waterway is busy. It is quite hard to find clear water.

Actually, the issue of which way the current is going is not clear cut. At times of heavy flood, river authorities routinely re-reverse the flow of the river to go back into the lake, in order to release pressure on the holding capacity of the river system. Taking water out of Lake Michigan instead of putting water into it over the course of a century has lowered the water level in the lake. The lower the lake level goes, the more the river strains to flowing back into it. Left to itself it would reverse back to its primordial direction of flow. Even when the bulk of the river is going the way it is humanly designed to flow, there are other parts going in the opposite direction at the same time. In an important study of flow in the Chicago River, bidirectional flow has been detected (Jackson et al. 2008). Minerals and polluting substances carried in solution find their own depth of suspension, with the heaviest near the bottom. Different layers of the river, being of different densities, react differently to the pull of gravity, thus traveling at different speeds in relation to each other. Thus the river has a kind of stratification. In this case, upper and middle layers go one way, deeper and denser layers go the other. There is overflow and underflow. The river is essentially trying to run in two directions at once.

As tour boats speed by, amplified voices giving historical information come down to us from their elevated decks, wailing in and out of earshot like spoken police car sirens. Muffled echoes of these amplified sounds bounce back from the concrete sides of the river. So do the rolling waves of the water. The best technique is to turn straight into the waves to avoid water swamping the canoe—then get ready for them to bounce back off the banks and hit the canoe from a different direction.

Specific narratives about the history of the river might seem clear and coherent to those on the tour boat deck, whose vantage point moves along with the source of the narration. But down near the surface of the water the arbitrary and foreshortened disembodied facts seem disconnected from the reality of the river itself. They get mixed together in incongruous ways. The water is choppy. So too is the acoustic and linguistic flowscape.

Because there is so much boat traffic we decide not to go all the way up to the river control structures (consisting essentially of a dam and lock) where the Main Stem meets Lake Michigan. Instead, we head back the way we came, past Wolf Point and up the North Branch.

It is worth noting that the Main Stem of the Chicago River is much longer today than it was 150 years ago. Not only is the stretch of river from Michigan Avenue to Lake Michigan entirely human-made, but so too is the ground on either side. It is all landfill. Rubble and charred debris from the Great Fire of Chicago in 1871 was dumped along the lake shoreline, and added to subsequently by spoil from excavation of basements and underground railways. As the reclaimed land pushed outward into the lake, forming what is now Grant Park, so the river lengthened accordingly.

Back on the North Branch, we take the river route around the west side of Goose Island this time. It is good to reach calm water again. There are no other boats on the river here.

FIGURE 12.5. The embanked river: (a) metal bank (b) concrete/earthen bank, with graffiti. Photographs by Jeffrey Benjamin.

Land on either side of the North Branch River is heavily industrialized. Goose Island is on the east bank. On the west bank is a salt packaging and warehouse facility. The vertical banks of the river are faced with retaining sheets of smooth or corrugated metal, allowing large barges bringing materials to dock and unload but making it impossible for a canoeist to pull a small boat into the shore and disembark. The old creeks and gullies and gently sloping banks have been replaced by vertical walls—metal in places, concrete elsewhere—which sharply separate the water from the land.

This is where something strange starts to happen—where our encounter with the river really begins. At first the signs are subtle, barely noticed. A slight difficulty in steering the boat. It is as though something is trying to spin the boat around, pushing the stern first this way, then that. It gets worse. Soon the movement of the boat becomes extremely erratic, almost out of control.

There are pros and cons of using small boats to explore rivers. Gone is the detached stance of an objective observer, and instead one assumes the more engaged attitude of an active participant. There is the sense of being in touch with the river and its flow. Being situated in the riverscape, you can act on the river and the river can act on you. Sometimes this develops into something like a wrestling match, with participants locked together in move and countermove. No true river encounter is possible without this interplay of human and river forces. Such an interplay necessarily includes within it the possibility that the river might exert its forces in undesired and unexpected ways.

Suddenly we are rising and falling on a series of large waves. There is a swell on the river estimated to be about 4 to 5 feet from peak to trough, with about 20 to 30 feet and 3 to 5 seconds from the peak of one wave to another—the kind of swell one might encounter on the sea or a large lake, whipped up by the wind, perhaps in the wake of a storm. It is difficult to tell which way it is going as it rebounds from one metal side to the other, almost breaking into turbulence where waves from different directions meet. We ride it out. It lasts a couple of minutes, and then the river reverts to its former calm state. What makes the experience so uncanny is that there is nothing at all to explain it. No boats. No wind. No impending storm. No sluices opening or closing on the side of the river. Nothing visible on the surface, upriver or down, that could have caused the river disturbance.

So far in this chapter there has been no mention of Deep Tunnel. It has been there throughout our journey, an unseen presence, about 200 feet below the riverbed. Now it is important to bring it into the discussion. Deep Tunnel is effectively an artificial underground river, concrete-lined, shadowing the course and gradient of the surface watercourse and connected to it by a network of interceptor tunnels, drop shafts, sewers, reservoirs, and pumping stations. Up to 30 feet in diameter, it extends for 110 miles in linked sections beneath the North Branch, Main Stream, South Branch, Calumet, and Des Plaines Rivers. It is one of the engineering marvels of the contemporary age.

The purpose of Deep Tunnel is to take the overflow of combined sewage and storm water from the city sewer and drainage systems that would otherwise empty into the river, to divert it into temporary holding reservoirs, to process it, pump it back to the surface, and return it into the river in a controlled manner (Scalise and Fitzpatrick 2012). The success of Deep Tunnel can be measured in the cleaner water of the Chicago River, though an unintended consequence (of removing the toxic barrier between watersheds) was to open up a possible gateway for invasive species to cross from one watershed to another, as discussed earlier.

The control structures of Deep Tunnel, regulating overflows and diversions of flows into and out of the river, are computerized. Technological systems and river are intermeshed. That makes it sound as though things are entirely under control, which is not always the case. In 1999 a powerful shock wave traveled the wrong way up the Main Stream section of the tunnel, causing immense amounts of damage. The wave surged downstream, rebounded at the end, then surged upstream, meeting itself on the way, strangely echoing the tendency of the surface river to try to go in both directions at once (Kendall 1999).

Combined storm water and sewage can travel the wrong way vertically as well as horizontally. In 1986 a full-to-capacity Deep Tunnel sent water surging up drop shafts like volcanic lava from the bowels of the earth, to erupt as geysers 65 feet high downtown, catapulting manhole covers into the air and flooding streets and basements with raw diluted sewage (Karwath 1986).

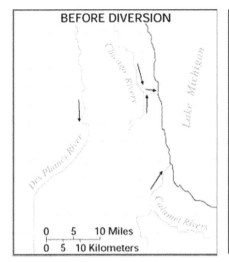
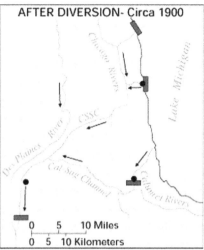

FIGURES 12.6A and 12.6B. Direction of flow of the Chicago River before after its reversal in 1900. Red blocks indicate positions of river control structures. United States Geological Survey.

We arrive back at the turning basin where we had set off, pulling the canoe back up onto the jetty in the knowledge that something significant has just occurred but not sure exactly what it was that we experienced, or how it could be explained.

It is difficult to say for sure, but Deep Tunnel probably had something to do with the river disturbance that we encountered. An inference drawn here is that when near full capacity, pressure within Deep Tunnel sends water backflowing up drop shafts into sewage interceptor tunnels and from there through automatic underwater control gates into the river. We just happened to be passing over such a gate when it opened. Alternatively, it may be that the drop shafts are closed off when Deep Tunnel is full, leaving nowhere for excess storm water in subsurface sewers to go except through outflow tunnels into the river. Both scenarios go some way toward explaining the hundreds of combined sewage outflows (CSOs) that take place each year, many of them on the North Branch, not all of them anticipated or recorded.

CONCLUSION

What kind of entity is the Chicago River? The dictionary definitions of rivers as natural watercourses flowing into an ocean or lake seem to fall hopelessly wide of the mark. This particular river is clearly much more than a natural flow of water; moreover it flows away from the lake it once flowed into.

One of the difficulties of describing rivers of the Anthropocene is finding the categories in which to place them, and the metaphors with which to describe them. The river in this case cannot be separated from control structures and river

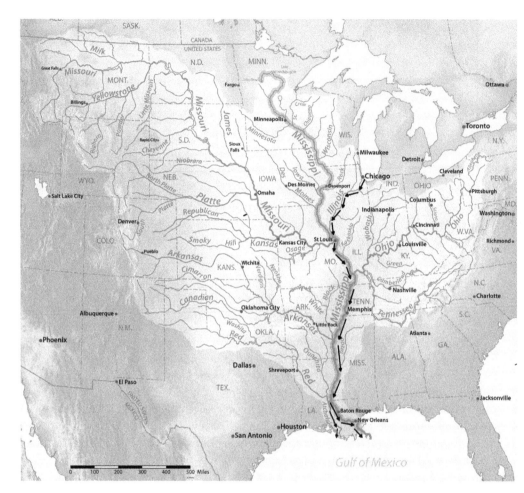

FIGURE 12.7. Flow of water (and sewage) from the Chicago River through other river systems into the Gulf of Mexico. Adapted from map by Shannon 1, CC by 4.0.

barriers, artificial embankments, concrete catchments, sewer and drainage networks, underground reservoirs and pumping stations, engineered flow regimes, and so on. From our encounter with it, there is a sense in which it is a partial manifestation of something much greater, the full extent of which we have not yet fully grasped. This is where Timothy Morton's concept of hyperobjects comes in. It stretches our concepts of time and space, challenging our notions of what objects are, while providing space for conceptualizing things that do not fit within usual frameworks. Hyperobjects are so large and multifaceted and spread out through time that they cannot be apprehended in one go, and they have aspects to them that may be hidden and inaccessible, phasing in and out of human awareness (Morton 2014).

The river might be considered part of something like that, which we just skimmed the surface of during our trip. For all that we have tried to formulate a phenomenological approach in this chapter, we are dealing with an entity that extends beyond perceived phenomena and the limits of embodied, situated perceptions in any given spatial and temporal context. It is larger than the river watershed, with arms or branches reaching as far away in space as the Gulf of Mexico. It is as tall as a city and has roots that go farther than one might think into the depths of the earth.

This is more than just a hybrid entity, organic machine, or cyborg.

It is a hyperobject.

BIBLIOGRAPHY

Aadland, Øyvind. 2002. "Sera: Traditionalism or Living Democratic Values? A Case Study on the Sidama in Southern Ethiopia." In *Ethiopia: The Challenge of Democracy from Below,* edited by B. Zewde and S. Pausewang, 29–44. Stockholm: Elanders Gotab.

Acreman, Mike, Angela H. Arthington, Matthew J. Colloff, Carol Couch, Neville D. Crossman, Fiona Dyer, Ian Overton, Carmel A. Pollino, Michael J. Stewardson, and William Young. 2014. "Environmental Flows for Natural, Hybrid, and Novel Riverine Ecosystems in a Changing World." *Frontiers in Ecology and the Environment* 12 (8): 466–73.

Acres International Limited (AIL). 1995a. "Feasibility Study of the Birr and Koga Irrigation Project. Koga Catchment and Irrigation Studies, Main Report." Transitional Government of Ethiopia, Ministry of Natural Resources and Environmental Protection, Water Resources Authority.

———. 1995b. Feasibility Study of the Birr and Koga Irrigation Project. Koga Catchment and Irrigation Studies, Main Report, Annexes M–O. Transitional Government of Ethiopia, Ministry of Natural Resources and Environmental Protection, Water Resources Authority, March.

Adam, Barbara. 1998. *Timescapes of Modernity: The Environment and Invisible Hazards.* Hove, U.K.: Psychology Press.

African Development Bank (AfDB). 2001. "Koga Irrigation and Watershed Management Project. Appraisal Report." African Development Bank, Abidjan.

Allen, Robert C. 2009. *The British Industrial Revolution in Global Perspective.* Cambridge: Cambridge University Press.

Anderson, Benedict. 1999. *Imagined Communities: Reflections on the Origin and Spread of Nationalism.* London: Verso.

Andersson, Andreas J., Fred T. Mackenzie, and Abraham Lerman. 2005. "Coastal Ocean and Carbonate Systems in the High CO_2 World of the Anthropocene." *American Journal of Science* 305 (9): 875–918.

Ansted, David Thomas. 1863. *The Great Stone Book of Nature.* London: Macmillan.

Antonaccio, Maria. 2017. "DeMoralizing and Remoralizing the Anthropocene." In *Religion in the Anthropocene*. Eugene, OR: Wipf and Stock.

Armesto, Juan J., Daniela Manuschevich, Alejandra Mora, C. Smith-Ramirez, R. Rozzi, A. M. Abarzúa, and P. A. Marquet. 2010. "From the Holocene to the Anthropocene: A Historical Framework for Land Cover Change in Southwestern South America in the Past 15,000 Years." *Land Use Policy* 27 (2): 148–60.

Arnaud-Fassetta, Gilles, Elsevier Masson, and Emmanuel Reynard. 2013. *European Continental Hydrosystems under Changing Water Policy*. Munich: Friedrich Pfeil Verlag. https://hal.archives-ouvertes.fr/hal-00878326.

Arsano, Yacob, and Imeru Tamrat. 2005. "Ethiopia and the Eastern Nile Basin." *Aquatic Sciences* 67 (1): 15–27.

Asimov, Isaac. 1952. "The Martian Way." *Galaxy Science Fiction*, November, 4–60.

Astbury, Arthur Kenelm. 1958. *The Black Fens*. Cambridge: Golden Head Press.

Baccini, Peter, and Paul H. Brunner. 2012. *Metabolism of the Anthroposphere: Analysis, Evaluation, Design*. Cambridge, MA: MIT Press.

Bairoch, Paul. "International Industrialization Levels from 1750 to 1980." *Journal of European Economic History* 11 (1982): 269–333.

Balthasar, Hans Urs von. 1988. *Theo-Drama: Theological Dramatic Theory*. Vol. 1: *Prolegomena*. Translated by Graham Harrison. San Francisco, CA: Ignatius Press.

Barles, Sabine. 1999. *La ville délétère: Médecins et ingénieurs dans l'espace urbain (XVIIIe-XXe siècles)*. Seyssel: Editions Champ Vallon.

———. 2005. *L'invention des déchets urbains: France (1790–1970)*. Seyssel: Editions Champ Vallon.

Barles, Sabine, and Jean-Marie Mouchel. 2006. *Man and River Systems: Interactions among Rivers, Their Watersheds, and the Sociosystem*. Paris: Presses de l'Ecole Nationale des Ponts et Chaussées.

Barnosky, Anthony D. 2014. "Palaeontological Evidence for Defining the Anthropocene." *Geological Society, London, Special Publications* 395 (1): 149–65.

Barnosky, Anthony D., Nicholas Matzke, Susumu Tomiya, Guinevere O. U. Wogan, Brian Swartz, Tiago B. Quental, Charles Marshall, et al. 2011. "Has the Earth/'s Sixth Mass Extinction Already Arrived?" *Nature* 471 (7336): 51–57.

Barrett, James. 2002. *Staged Narrative: Poetics and the Messenger in Greek Tragedy*. Berkeley: University of California Press.

Basso, Keith H. 1996. *Wisdom Sits in Places: Landscape and Language among the Western Apache*. Albuquerque: University of New Mexico Press.

Bates, Bryson, Zbigniew W. Kundzewicz, Shaohong Wu, and Jean Palutikof. 2008. *Climate Change and Water: Technical Paper vi*. Geneva, Swizerland: Intergovernmental Panel on Climate Change (IPCC). http://digital.library.unt.edu/ark:/67531/metadc11958/m1/13/.

Beinart, William J., and Lotte Hughes. 2007. "Forests and Forestry in India." In *Environment and Empire*, 111–29. Oxford History of the British Empire Companion. Oxford: Oxford University Press.

Bendjoudi, Hocine, P. Weng, Roger Guérin, and J. F. Pastre. 2002. "Riparian Wetlands of the Middle Reach of the Seine River (France): Historical Development, Investigation and Present Hydrologic Functioning. A Case Study." *Journal of Hydrology* 263 (1): 131–55.

Benlloch, Ana, Nikki Pugh, and Stuart Tait. 2008. "The Ludogeographic Society." http://ludogeography.npugh.co.uk. Accessed 17 October 2016.

Bennett, Jane. 2009. *Vibrant Matter: A Political Ecology of Things.* Durham, NC: Duke University Press.

Berner, Elizabeth Kay, and Robert A. Berner. 1996. *Global Environment: Water, Air, and Geochemical Cycles.* Princeton, NJ: Princeton University Press.

Biermann, F., K. Abbott, S. Andresen, K. Bäckstrand, S. Bernstein, M. M. Betsill, H. Bulkeley, et al. 2012. "Navigating the Anthropocene: Improving Earth System Governance." *Science* 335 (6074): 1306–7.

Billen, Gilles, Sabine Barles, Petros Chatzimpiros, and Josette Garnier. 2012. "Grain, Meat and Vegetables to Feed Paris: Where Did and Do They Come From? Localising Paris Food Supply Areas from the Eighteenth to the Twenty-First Century." *Regional Environmental Change* 12 (2): 325–35.

Billen, Gilles, Sabine Barles, Josette Garnier, Joséphine Rouillard, and Paul Benoit. 2009. "The Food-Print of Paris: Long-Term Reconstruction of the Nitrogen Flows Imported into the City from Its Rural Hinterland." *Regional Environmental Change* 9 (1): 13–24.

Billen, Gilles, Josette Garnier, Jean-Marie Mouchel, and Marie Silvestre. 2007. "The Seine System: Introduction to a Multidisciplinary Approach of the Functioning of a Regional River System." *Science of the Total Environment* 375 (1–3): 1–12.

Blaikie, Piers, and Harold Brookfield. 2015. *Land Degradation and Society.* London: Routledge.

Bloor, David. 1976. "The Strong Programme in the Sociology of Knowledge." In *Knowledge and Social Imagery,* 1–19. Chicago: University of Chicago Press.

Boët, Philippe, Jérôme Belliard, Romuald Berrebi-dit-Thomas, and Evelyne Tales. 1999. "Multiple Human Impacts by the City of Paris on Fish Communities in the Seine River Basin, France." In *Man and River Systems: The Functioning of River Systems at the Basin Scale,* edited by J. Garnier and J.-M. Mouchel, 59–68. Developments in Hydrobiology. Dordrecht: Springer.

Bradshaw, A. D. 1988. "Alternate Endpoints for Restoration." In *Rehabilitating Damaged Ecosystems,* edited by J. J. Cairns, 2:69–85. Boca Raton, FL: CRC Press.

Braje, Todd J., and Jon M. Erlandson. 2013a. "Human Acceleration of Animal and Plant Extinctions: A Late Pleistocene, Holocene, and Anthropocene Continuum." *Anthropocene, When Humans Dominated the Earth: Archeological Perspectives on the Anthropocene* 4 (December): 14–23.

———. 2013b. "Looking Forward, Looking Back: Humans, Anthropogenic Change, and the Anthropocene." *Anthropocene, When Humans Dominated the Earth: Archeological Perspectives on the Anthropocene* 4 (December): 116–21.

Braungart, Michael, and William McDonough. 2009. *Cradle to Cradle.* London: Vintage.

Brigden, Kevin, Iryna Labunska, Paul Johnston, and David Santillo. 2012. "Organic Chemical and Heavy Metal Contaminants from Communal Wastewater Treatment Plants with Links to Textile Manufacturing, and in River Water Impacted by Wastewater from a Textile Dye Manufacturing Facility, in China." Greenpeace Research Laboratories Technical Report 07/2012. Greenpeace Research Laboratories, University of Exeter.

BRL Ingénierie (BRLI). 2009. "Consultancy Services to Prepare Regulation for Agricultural Water Users' Associations: First Draft Model Agreements." Ethiopian Nile Irrigation and Drainage Project."

Brown, Antony G., Stephen Tooth, Joanna E. Bullard, David S. G. Thomas, Richard C. Chiverrell, Andrew J. Plater, Julian Murton, et al. 2016. "The Geomorphology of the Anthropocene: Emergence, Status and Implications." *Earth Surface Processes and Landforms* 42 (1): 71–90.

Brown, Claire, Matt Walpole, Lucy Simpson, and Megan Tierney. 2016. "Introduction to the UK National Ecosystem Assessment," October. http://citeseerx.ist.psu.edu/viewdoc/download?doi=10.1.1.692.7181&rep=rep1&type=pdf.

Brown, James Robert. 2004. *Who Rules in Science? An Opinionated Guide to the Wars.* Cambridge, MA: Harvard University Press.

Brown's South Africa: A Practical and Complete Guide for the Use of Tourists, Sportsmen, Invalids, and Settlers. 1893. London: Sampson Low, Marston and Company.

Bryant, Raymond L., and Sinéad Bailey. 1997. *Third World Political Ecology.* London: Routledge.

Butler, Graham. 2012. "Disease, Medicine and the Urban Poor in Newcastle-upon-Tyne, C. 1750–1850." PhD dissertation, Newcastle University, Newcastle, U.K. https://theses.ncl.ac.uk/dspace/bitstream/10443/1501/1/Butler%2012.pdf.

Cadiz, Charles Fitzwilliam, and Robert Lyon. 1891. "'Law No. 5, 1859.'" In *Natal Ordinances, Laws, and Proclamations: Compiled and Edited Under the Authority and with the Sanction of His Excellency the Lieutenant Govenor and the Honorable the Legislative Council.* Vol. 1. Cape Town: W. M. Watson.

Cairns, John, Jr. 1972. "Rationalization of Multiple Use of Rivers." In *River Ecology and Man,* edited by Ray Oglesby, Clarence Carlson, and James McCann, 421–30. New York: Academic Press.

Camillus, John. 2008. "Strategy as a Wicked Problem." *Harvard Business Review* 86 (5): 99–106.

Carey, Mark. 2012. "Climate and History: A Critical Review of Historical Climatology and Climate Change Historiography." *Wiley Interdisciplinary Reviews: Climate Change* 3 (3): 233–49.

Carpenter, Stephen R., Stuart G. Fisher, Nancy B. Grimm, and James F. Kitchell. 1992. "Global Change and Freshwater Ecosystems." *Annual Review of Ecology and Systematics* 23: 119–39.

Carpenter, Stephen R., Harold A. Mooney, John Agard, Doris Capistrano, Ruth S. DeFries, Sandra Díaz, Thomas Dietz, et al. 2009. "Science for Managing Ecosystem Services: Beyond the Millennium Ecosystem Assessment." *Proceedings of the National Academy of Sciences* 106 (5): 1305–12.

Carpenter, Stephen R., Emily H. Stanley, and M. Jake Vander Zanden. 2011. "State of the World's Freshwater Ecosystems: Physical, Chemical, and Biological Changes." *Annual Review of Environment and Resources* 36 (1): 75–99.

Carson, Rachel. [1962] 2002. *Silent Spring.* Boston: Houghton Mifflin Harcourt.

Cascão, Ana Elisa. 2008. "Ethiopia: Challenges to Egyptian Hegemony in the Nile Basin." *Water Policy* 10 (2): 13–28.

Castonguay, Stéphane, and Matthew D. Evenden, eds. 2012. *Urban Rivers: Remaking Rivers, Cities, and Space in Europe and North America*. History of the Urban Environment. Pittsburgh, PA: University of Pittsburgh Press.

Cetina, Karin Knorr. 1999. *Epistemic Cultures: How the Sciences Make Knowledge*. Cambridge, MA: Harvard University Press.

Chakrabarty, Dipesh. 2009. "The Climate of History: Four Theses." *Critical Inquiry* 35 (2): 197–222.

———. 2012. "Postcolonial Studies and the Challenge of Climate Change." *New Literary History* 43 (1): 1–18.

Clarke, Arthur Charles. 1951. *The Sands of Mars*. London: Sidgwick and Jackson.

Coates, Peter. 2013. *A Story of Six Rivers: History, Culture and Ecology*. London: Reaktion Books.

Coillard, François. 1897. *On the Threshold of Central Africa: A Record of Twenty Years' Pioneering among the Barotsi of the Upper Zambesi*. Translated by Catherine Winkworth Mackintosh. London: Hodder and Stoughton.

Cooper, Sherri Rumer, and Grace S. Brush. 1993. "A 2,500-Year History of Anoxia and Eutrophication in Chesapeake Bay." *Estuaries* 16 (3): 617–26.

Corfield, P. J. 2007. *Time and the Shape of History*. New Haven, CT: Yale University Press.

Costanza, Robert, Lisa Graumlich, and William L. Steffen. 2007. *Sustainability Or Collapse? An Integrated History and Future of People on Earth*. Cambridge, MA: MIT Press.

Costanza, Robert, Bobbi S. Low, Elinor Ostrom, and James Wilson. 2001. *Institutions, Ecosystems, and Sustainability*. London: Lewis Publishers.

Costanza, Robert, Sander van der Leeuw, Kathy Hibbard, Steve Aulenbach, Simon Brewer, Michael Burek, Sarah Cornell, et al. 2012. "Developing an Integrated History and Future of People on Earth (IHOPE)." *Current Opinion in Environmental Sustainability*, Open issue, 4 (1): 106–14.

Crosby, Alfred W., Jr. 1973. *The Columbian Exchange: Biological and Cultural Consequences of 1492*. Westport, CT: Greenwood Press.

———. [1986] 2004. *Ecological Imperialism: The Biological Expansion of Europe, 900–1900*. 2nd ed. Cambridge: Cambridge University Press.

Crossland, Christopher J., Hartwig H. Kremer, Han Lindeboom, Janet I. Marshall Crossland, and Martin D. A. Le Tissier. 2006. *Coastal Fluxes in the Anthropocene: The Land-Ocean Interactions in the Coastal Zone Project of the International Geosphere-Biosphere Programme*. Berlin: Springer.

Crutzen, Paul. 1970. "The Influence of Nitrogen Oxides on the Atmospheric Ozone Content." *Quarterly Journal of the Royal Meteorological Society* 96 (408): 320–25.

———. 1974. "Estimates of Possible Variations in Total Ozone Due to Natural Causes and Human Activities." *Ambio* 3 (6): 201–10.

———. 2002. "Geology of Mankind." *Nature* 415 (6867): 23–23.

Crutzen, Paul, and Dieter Ehhalt. 1977. "Effects of Nitrogen Fertilizers and Combustion on the Stratospheric Ozone Layer." *Ambio* 6 (2–3): 112–17.

Crutzen, Paul J., and Will Steffen. 2016. "How Long Have We Been in the Anthropocene Era?" *Climatic Change* 61 (3): 251–57.

Crutzen, Paul J., and Eugene Stoermer. 2000a. "The Anthropocene." *Global Change Newsletter* 41 (1): 17–18.

———. 2000b. "The Anthropocene IGBP Newsletter, 41." *Royal Swedish Academy of Sciences, Stockholm, Sweden.*

Daily, Gretchen, and Katherine Ellison. 2003. *The New Economy of Nature: The Quest to Make Conservation Profitable.* Washington, DC: Island Press.

Daily, Gretchen C., Peter M. Kareiva, Stephen Polasky, Taylor H. Ricketts, and Heather Tallis, eds. 2011. *Natural Capital: Theory & Practice of Mapping Ecosystem Services.* Oxford: Oxford University Press.

Dalby, Simon. 2007. "Anthropocene Geopolitics: Globalisation, Empire, Environment and Critique." *Geography Compass* 1 (1): 103–18.

Darby, Henry Clifford. 1983. *Changing Fenland.* Cambridge: Cambridge University Press.

Das, Pallavi V. 2005. "Hugh Cleghorn and Forest Conservancy in India." *Environment and History* 11 (1): 55–82.

Daston, Lorraine J., and Peter Galison. 1992. "The Image of Objectivity." *Representations,* no. 40: 81–128.

———. 2010. *Objectivity.* Brooklyn, NY: Zone Books.

Davies, Matthew I. J., and Freda Nkirote M'Mbogori. 2013. *Humans and the Environment: New Archaeological Perspectives for the Twenty-First Century.* Oxford: Oxford University Press.

Davies, Norman. 2012. *Vanished Kingdoms: The History of Half-Forgotten Europe.* London: Penguin.

Davis, Robert. 2011. "Inventing the Present: Historical Roots of the Anthropocene." *Earth Sciences History* 30 (1): 63–84.

Davy, Humphry. 1830. *Consolations in Travel, or the Last Days of a Philosopher.* London.

De Vos, Jurriaan M., Lucas N. Joppa, John L. Gittleman, Patrick R. Stephens, and Stuart L. Pimm. 2015. "Estimating the Normal Background Rate of Species Extinction." *Conservation Biology: The Journal of the Society for Conservation Biology* 29 (2): 452–62.

Dean, Jonathan R., Melanie J. Leng, and Anson W. Mackay. 2014. "Is There an Isotopic Signature of the Anthropocene?" *Anthropocene Review* (July): 2053019614541631.

Deane-Drummond, Celia. 2004. *The Ethics of Nature.* Malden, MA: Blackwell.

———. 2010. "Beyond Humanity's End: An Exploration of a Dramatic versus Narrative Rhetoric and Its Ethical Implications." In *Future Ethics: Climate Change and Apocalyptic Imagination,* edited by Stefan Skrimshire, 242–59. London: Continuum.

———. 2014. *The Wisdom of the Liminal: Evolution and Other Animals in Human Becoming.* Grand Rapids, MI: Wm. B. Eerdmans.

Department for Environment, Food, and Rural Affairs (DEFRA). 2013. "Catchment Based Approach: Improving the Quality of Our Water Environment—Publications." London: DEFRA. https://www.gov.uk/government/publications/catchment-based-approach-improving-the-quality-of-our-water-environment.

Dewulf, Art, Greet François, Claudia Pahl-Wostl, and Tharsi Taillieu. 2007. "A Framing Approach to Cross-Disciplinary Research Collaboration: Experiences from a Large-Scale Research Project on Adaptive Water Management." *Ecology and Society* 12 (2): 14.

Diamond, Jared. 2005. *Collapse: How Societies Choose to Fail or Succeed.* New York: Penguin.

Dibley, Ben. 2012. "'The Shape of Things to Come': Seven Theses on the Anthropocene and Attachment." *Australian Humanities Review* 52: 139–53.

Dobbs, Stephen. 2002. "Urban Redevelopment and the Forced Eviction of Lighters from the Singapore River." *Singapore Journal of Tropical Geography* 23 (3): 288–310.

Donald, Thomas. 1774. *Thomas Donald Historic Map of Cumberland 1774: Reproduced in Six Sections from the Original Engravings.* Carlisle: Cumberland and Westmorland Antiquarian and Archaeological Society.

Douglas, Ian, and Nigel Lawson. 2000. "The Human Dimensions of Geomorphological Work in Britain." *Journal of Industrial Ecology* 4 (2): 9–33.

Dove, Michael R., and Carol Carpenter, eds. 2008. *Environmental Anthropology. A Historical Reader.* Oxford: Blackwell.

———. 2013. "Environmental Anthropology: A Historical Reader." SSRN Scholarly Paper ID 2295083. Social Science Research Network, Rochester, NY. https://papers.ssrn.com/abstract=2295083.

Downs, Peter, and Ken Gregory. 2014. *River Channel Management: Towards Sustainable Catchment Hydrosystems.* Routledge.

Dubord, Guy. 1955. "Introduction to a Critique of Urban Geography." Translated by Ken Knabb. Situationist International Online. http://www.cddc.vt.edu/sionline/presitu/geography.html. Accessed 17 October 2016. Originally published as "Introduction à une critique de la geographie urbaine" in *Les Lèvres Nues,* no. 6 (September 1955).

Dubos, Rene. 1980. *The Wooing of Earth: New Perspectives on Man's Use of Nature.* New York: Scribner's.

Duchesne, Ricardo. 2004. "On the Rise of the West: Researching Kenneth Pomeranz's Great Divergence." *Review of Radical Political Economics* 36 (1): 52–81.

Edgeworth, Matt. 2011. *Fluid Pasts: Archaeology of Flow.* London: Bloomsbury Academic Press.

Egan, Michael. 2002. "Subaltern Environmentalism in the United States: A Historiographic Review." *Environment and History* 8 (1): 21–41.

Ehrlich, Paul R. 1968. *The Population Bomb: Population Control or Race to Oblivion.* New York: Ballantine Books.

Elkins, James W. 1999. "Chlorofluorocarbons (CFCs)." In *The Chapman & Hall Encyclopedia of Environmental Science,* edited by David E. Alexander and Rhodes W. Fairbridge. Boston: Kluwer Academic.

Ellis, Erle C., Dorian Q. Fuller, Jed O. Kaplan, and Wayne G. Lutters. 2013. "Dating the Anthropocene: Towards an Empirical Global History of Human Transformation of the Terrestrial Biosphere." *Elementa: Science of the Anthropocene* 1 (December): 18.

Ellis, Erle C., Jed O. Kaplan, Dorian Q. Fuller, Steve Vavrus, Kees Klein Goldewijk, and Peter H. Verburg. 2013. "Used Planet: A Global History." *Proceedings of the National Academy of Sciences* 110 (20): 7978–85.

Ellis, Erle C., and Navin Ramankutty. 2008. "Putting People in the Map: Anthropogenic Biomes of the World." *Frontiers in Ecology and the Environment* 6 (8): 439–47.

Environment Agency. 2000. "Local Environment Action Plan, Northumbrian Region." Environment Agency, Newcastle-upon-Tyne.

Erkens, Gilles, Michiel ven der Meulen, and Hans Middelkoop. 2016. "Double Trouble: Subsidence and CO_2 Respiration due to 1,000 Years of Dutch Coastal Peatlands Cultivation." *Hydrogeology Journal* 24 (3): 551–68.

Escobar, Arturo. 2008. *Territories of Difference: Place, Movements, Life, Redes*. Durham, NC: Duke University Press.

"Ethiopia—Irrigation and Drainage Project." 2007. 39866. World Bank. http://documents.worldbank.org/curated/en/454731468256452428/Ethiopia-Irrigation-and-Drainage-Project.

European Commission. 2013. "Science for Environment Policy In-Depth Report: Environmental Citizen Science." European Commission DG Environment. http://ec.europa.eu/science-environment-policy.

European Space Agency. n.d. "Paul J. Crutzen: The Engineer and the Ozone Hole." http://www.esa.int/About_Us/Welcome_to_ESA/ESA_history/Paul_J._Crutzen_The_engineer_and_the_ozone_hole.

Everard, Mark. 2012. "Why Does 'Good Ecological Status' Matter?" *Water and Environment Journal* 26 (2): 165–74.

Everard, Mark, and Robert McInnes. 2013. "Systemic Solutions for Multi-Benefit Water and Environmental Management." *Science of the Total Environment* 461–62 (September): 170–79.

Farman, Joseph C. 1987. "Recent Measurements of Total Ozone at British Antarctic Survey Stations." *Philosophical Transactions of the Royal Society of London A: Mathematical, Physical and Engineering Sciences* 323 (1575): 629–44.

Farman, Joseph C., Brian G. Gardiner, and Jonathan D. Shanklin. 1985. "Large Losses of Total Ozone in Antarctica Reveal Seasonal ClOx/NOx Interaction." *Nature* 315 (6016): 207–10.

Farrell, Katharine N., and Andreas Thiel. 2013. "Nudging Evolution." *Ecology and Society* 18 (4): 47.

Feyerabend, Paul. 1975. *Against Method: Outline of an Anarchistic Theory of Knowledge*. London: Humanities Press.

Floud, Roderick, and Paul Johnson, eds. 2004. *The Cambridge Economic History of Modern Britain*. Vol. 1. Cambridge: Cambridge University Press.

Fogg, Martyn J. 1995. *Terraforming: Engineering Planetary Environments*. Warrendale, PA: Society of Automotive Engineers.

Folke, Carl. 2006. "Resilience: The Emergence of a Perspective for Social–Ecological Systems Analyses." *Global Environmental Change* 16 (3): 253–67.

Food and Agriculture Organization of the United Nations (FAO). 2001. "Overview Paper: Irrigation Management Transfer, Sharing Lessons from Global Experience. International E-Mail Conference on Irrigation Management Transfer." http://www.fao.org/nr/water/docs/irrigation/Overview.pdf.

"The Forests of Pegu." 1856. *Allen's Indian Mail*, April. about:home.

Frank, Andre Gunder. 1998. *ReORIENT: Global Economy in the Asian Age*. 1st ed. Berkeley: University of California Press.

French, Charles A. I. 2003. *Geoarchaeology in Action: Studies in Soil Micromorphology and Landscape Evolution*. London: Routledge.

Frenken, Karen. 2005. *Irrigation in Africa in Figures: AQUASTAT Survey, 2005*. Vol. 29. Rome, Italy: Food and Agriculture Organization.

Gadgil, Madhav, and Ramachandra Guha. 1993. *This Fissured Land: An Ecological History of India*. Berkeley: University of California Press.

Gaffney, Owen. 2009. "A Planet on the Edge—IGBP." *Global Change–IGBP.* http://www. igbp.net/news/features/features/aplanetontheedge.5.1b8ae20512db692f2a680003122. html.

Garnier, Josette, Gilles Billen, and Aurélie Cébron. 2007. "Modelling Nitrogen Transformations in the Lower Seine River and Estuary (France): Impact of Wastewater Release on Oxygenation and N$_2$O Emission." *Hydrobiologia* 588 (1): 291–302.

Garnier, Josette, Luis Lassaletta, Gilles Billen, Estela Romero, Bruna Grizzetti, Julien Némery, Thi Phuong Quynh Le, et al. 2015. "Phosphorus Budget in the Water-Agro-Food System at Nested Scales in Two Contrasted Regions of the World (ASEAN-8 and EU-27)." *Global Biogeochemical Cycles* 29 (9): 2015GB005147.

Garnier, Josselin, and Jean-Marie Mouchel, eds. 1999. *Man and River Systems: The Functioning of River Systems at the Basin Scale.* Dordrecht: Springer Science & Business Media.

Garrels, Robert M., Fred T. Mackenzie, and Cynthia Hunt. 1975. *Chemical Cycles and the Global Environment: Assessing Human Influences.* Los Altos, CA: W. Kaufmann.

Garret, William. [1818] 2010. *An Account of the Great Floods in the Rivers Tyne, Tees, Wear, Eden, Etc. in 1771 and 1815.* Whitefish, MT: Kessinger Publishing.

Gebeyehu, Admasu. 2004. "The Role of Large Water Reservoirs." In *Proceedings of 2nd International Conference on the Ethiopian Economy,* 14–16. Addis Ababa: Ethiopian Economic Association.

Gebre, Ayalew, Derese Getachew, and Matthew McCartney. 2007. "Stakeholder Analysis of the Koga Irrigation and Watershed Management Project." Report for the International Water Management Institute. http://publications.iwmi.org/pdf/H040845.pdf.

Geertz, Clifford. 1973. *The Interpretation of Cultures.* New York: Basic Books.

Godwin, Harry. 1978. *Fenland: Its Ancient Past and Uncertain Future.* Cambridge: Cambridge University Press.

Goffman, Erving. 1959. *The Presentation of Self in Everyday Life.* 1st ed. New York: Anchor Books.

Golinski, Jan. 2005. *Making Natural Knowledge: Constructivism and the History of Science.* Chicago: University of Chicago Press.

Graf, William L. 1999. "Dam Nation: A Geographic Census of American Dams and Their Large-Scale Hydrologic Impacts." *Water Resources Research* 35 (4): 1305–11.

Graham, Stephen. 2010. *Disrupted Cities: When Infrastructure Fails.* New York: Routledge.

Graham, Stephen, and Colin McFarlane. 2015. Introduction to *Infrastructural Lives: Urban Infrastructure in Context,* edited by Stephen Graham and Colin McFarlane, 1–14. New York: Routledge.

Groenfeldt, David, and Ashok Subramanian. 1998. *Handbook On Participatory, Irrigation, Management.* Washington, DC: Economic Development Institute of the World Bank. http://library.wur.nl/WebQuery/clc/976594.

Gross, Paul R., and Norman Levitt. 1994. *Higher Superstition: The Academic Left and Its Quarrels with Science.* Baltimore, MD: Johns Hopkins University Press.

Grove, Richard H. 1996. *Green Imperialism: Colonial Expansion, Tropical Island Edens and the Origins of Environmentalism, 1600–1860.* Cambridge: Cambridge University Press.

Grumbine, R. Edward. 1994. "What Is Ecosystem Management?" *Conservation Biology* 8 (1): 27–38.

Guldi, Jo, and David Armitage. 2014. *The History Manifesto.* Cambridge: Cambridge University Press.

Gupta, Avijit. 1982. "Observations on the Effects of Urbanization on Runoff and Sediment Production in Singapore." *Singapore Journal of Tropical Geography* 3 (2): 137–46.

Hacking, Ian. 2000. *The Social Construction of What?* Cambridge, MA: Harvard University Press.

Haileslassie, Amare, Fitsum Hagos, Everisto Mapedza, Claudia Sadoff, Seleshi Bekele Awulachew, Solomon Gebreselassie, and Don Peden. 2009. *Institutional Settings and Livelihood Strategies in the Blue Nile Basin: Implications for Upstream/Downstream Linkages.* Colombo, Sri Lanka: IWMI.

Haines-Young, Roy, and Marion Potschin. 2010. "The Links between Biodiversity, Ecosystem Services and Human Well-Being." In *Ecosystem Ecology: A New Synthesis,* edited by David G. Raffaeli and Christopher L. J. Frid, 110–39. Ecological Reviews. Cambridge: Cambridge University Press.

Hall, David. 1996. *The Fenland Project, Number 10: Cambridgeshire Survey, the Isle of Ely and Wisbech.* East Anglian Archaeology 79. Ipswich, U.K.: Cambridgeshire Archaeological Committee in conjunction with the Fenland Project Committee and the Scole Archaeological Committee. http://www.openbibart.fr/item/display/10068/928388.

Halliwell, Stephen. 1986. *Aristotle's Poetics.* 1st ed. Chapel Hill: University of North Carolina Press.

"Haltwhistle Burn—Newcastle University." 2016. October 19. http://research.ncl.ac.uk/haltwhistleburn/.

Hamilton, Clive, and Jacques Grinevald. 2015. "Was the Anthropocene Anticipated?" *Anthropocene Review* 2 (1): 59–72.

Hannon, Bruce. 1994. "Sense of Place: Geographic Discounting by People, Animals and Plants." *Ecological Economics* 10 (2): 157–74.

Haraway, Donna. 1988. "Situated Knowledges: The Science Question in Feminism and the Privilege of Partial Perspective." *Feminist Studies* 14 (3): 575–99.

———. 2015. "Anthropocene, Capitalocene, Plantationocene, Chthulucene: Making Kin." *Environmental Humanities* 6 (1): 159–65.

Hardin, Garrett. 1968. "The Tragedy of the Commons." *Science* 162 (3859): 1243–48.

Harrigan, S., C. Murphy, J. Hall, R. L. Wilby, and J. Sweeney. 2013. "Attribution of Detected Changes in Streamflow Using Multiple Working Hypotheses." *Hydrology and Earth System Sciences Discussions* 10 (10): 12373–416.

Hassan, Fekri. 2011. *Water History for Our Times.* Vol. 2. IHP Essays on Water History. Paris: UNESCO.

Hay, Carling C., Eric Morrow, Robert E. Kopp, and Jerry X. Mitrovica. 2015. "Probabilistic Reanalysis of Twentieth-Century Sea-Level Rise." *Nature* 517 (7535): 481–84. doi:10.1038/nature14093.

Heinlein, Robert A. 1950. *Farmer in the Sky.* New York: Scribner's.

Hinterthuer, Adam. 2012. "The Explosive Spread of Asian Carp: Can the Great Lakes Be Protected? Does It Matter?" *BioScience* 62 (3): 220–24.

Hohner, Susan M., and Thomas W. Dreschel. 2015. "Everglades Peats: Using Historical and Recent Data to Estimate Predrainage and Current Volumes, Masses and Carbon Contents." *Mires and Peat* 16: 1–15.

Holling, Crawford Stanley. 1978. *Adaptive Environmental Assessment and Management.* Chichester, U.K.: Wiley-InterScience.

Homer-Dixon, Thomas F. 1994. "Environmental Scarcities and Violent Conflict: Evidence from Cases." *International Security* 19 (1): 5–40.

Hornborg, Alf, and Carole L. Crumley, eds. 2006. *The World System and the Earth System: Global Socioenvironmental Change and Sustainability since the Neolithic.* Walnut Creek, CA: Left Coast Press.

Horton, A. 1989. *Geology of the Peterborough District: Memoir for 1: 50,000 Geological Sheet 158.* London: British Geological Survey.

House of Commons, Great Britain. 1906. "Law (Enacted by the Governor of Natal, with the Advice and Consent of the Legislative Council Thereof) 'To Make the Provision for the Better Preservation of Game in the Colony of Natal.'" In *Parliamentary Accounts and Papers: Colonies and British Possessions—Continued. Africa. Session 13 February 1906–21 December 1906.* Vol. 15. London: H. M. Stationery Office.

Hudson, Mark James. 2014. "Dark Artifacts: Hyperobjects and the Archaeology of the Anthropocene." *Journal of Contemporary Archaeology* 1 (1): 82–86.

Institute of Public and Environmental Affairs, Friends of Nature, Green Beagle, Environmental Protection Commonwealth Association, and Nanjing Green Stone Environmental Action Network. 2012. "Green Choice Apparel Supply Chain Investigation." http://www.ipe.org.cn/Upload/Report-Textiles-One-EN.pdf.

Intergovernmental Panel on Climate Change (IPCC). 2013a. "Climate Change 2013: The Physical Science Basis." IPCC. http://www.climatechange2013.org/report/.

———. 2013b. "Climate Change 2013: The Physical Science Basis, Summary for Policy Makers." IPCC. http://www.climatechange2013.org/report/.

———. 2014. "Climate Change 2014: Synthesis Report Summary for Policy Makers." IPCC. https://www.ipcc.ch/pdf/assessment-report/ar5/syr/AR5_SYR_FINAL_SPM.pdf.

Jackson, P. Ryan, Carlos M. García, Kevin A. Oberg, Kevin K. Johnson, and Marcelo H. García. 2008. "Density Currents in the Chicago River: Characterization, Effects on Water Quality, and Potential Sources." *Science of the Total Environment* 401 (1–3): 130–43.

Jäger, Jill, Gísili Pálsson, Michael Goodsite, Claudia Pahl-Wostl, Karen O'Brien, Leen Hordijk, Bernard Avril, et al. 2012. "Responses to Environmental and Societal Challenges for Our Unstable Earth (RESCUE), ESF Forward Look—ESF-COST 'Frontier of Science' Joint Initiative. European Science Foundation, Strasbourg and European Cooperation in Science and Technology, Brussels." http://www.esf.org/fileadmin/Public_documents/Publications/rescue.pdf.

Jahn, Thomas, Matthias Bergmann, and Florian Keil. 2012. "Transdisciplinarity: Between Mainstreaming and Marginalization." *Ecological Economics* 79: 1–10.

Jevons, William Stanley. 1906. *The Coal Question: An Inquiry Concerning the Progress of the Nation, and the Probable Exhaustion of Our Coal-Mines.* Reprint. New York: Economic Classics.

Jones, Jeffrey M. 2014. "In U.S., Most Do Not See Global Warming as Serious Threat." *Gallup.com.* March 13. http://www.gallup.com/poll/167879/not-global-warming-serious-threat.aspx.

Jones, P. D., D. H. Lister, and E. Kostopoulou. 2004. "Reconstructed River Flow Series from 1860s to Present." Science Report SC040052/SR. Bristol, U.K.: Environment Agency (UK).

Jong, Irene de. 1991. *Narrative in Drama: The Art of the Euripidean Messenger-Speech.* Leiden: E. J. Brill.

Jonsson, Fredrik Albritton. 2012. "The Industrial Revolution in the Anthropocene." *Journal of Modern History* 84 (3): 679–96.

Joyce, Patrick. 2010. "Filing the Raj: Political Technologies of the Imperial British State." In *Material Powers: Cultural Studies, History, and the Material Turn,* 151–53. New York: Routledge.

Kane, Stephanie C. 2009. "Stencil Graffiti in Urban Waterscapes of Buenos Aires and Rosario, Argentina." *Crime, Media, Culture* 5 (1): 9–28.

———. 2012. *Where Rivers Meet the Sea.* Philadelphia, PA: Temple University Press.

———. 2017. "Enclave Ecology: Hardening the Land-Sea Edge to Provide Freshwater in Singapore's Hydrohub." *Human Organization* 76 (1): 82–95.

———. Forthcoming. "Where Sheets of Water Intersect: Infrastructural Culture from Flooding to Hydropower in Winnipeg, Manitoba." In *Territory beyond Terra,* edited by Kimberley Peters, Philip Steinberg, and Elaine Stratford. Lanham, MD: Rowman & Littlefield International.

Kane, Stephanie C., Eden Medina, and Daniel M. Michler. 2015. "Infrastructural Drift in Seismic Cities Chile, Pacific Rim, 27 February 2010." *Social Text* 33 (1 122): 71–92.

Karr, James R., and Ellen W. Chu. 1999. *Restoring Life in Running Waters: Better Biological Monitoring.* Washington, D.C.: Island Press.

Karwath, Rob. 1986. "Tunnel's Geyser Effect Still Puzzling." *Chicago Tribue,* October 17, sec. Chicagoland.

Kelly, Jason. 2014. "The Anthropocene and Transdisciplinarity." *Journal of Contemporary Archaeology* 1 (1): 91–96.

Kendall, P. 1999. "Deep Tunnel Sealed Just in Time." *Chicago Tribune,* June 12.

Kendon, Elizabeth J., Nigel M. Roberts, Hayley J. Fowler, Malcolm J. Roberts, Steven C. Chan, and Catherine A. Senior. 2014. "Heavier Summer Downpours with Climate Change Revealed by Weather Forecast Resolution Model." *Nature Climate Change* 4 (7): 570–76.

Kim, Ungtae. 2008. *Climate Change Impacts on Hydrology and Water Resources of the Upper Blue Nile River Basin, Ethiopia.* Colombo, Sri Lanka: IWMI.

King, Anthony D. 1990. *Urbanism, Colonialism, and the World-Economy: Cultural and Spatial Foundations of the World Urban System.* London: Routledge.

Kirch, Patrick V. 2005. "Archaeology and Global Change: The Holocene Record." *Annual Review of Environment and Resources* 30 (1): 409–40.

Kloos, Helmut, and Worku Legesse. 2010. *Water Resources Management in Ethiopia: Implications for the Nile Basin.* Amherst, NY: Cambria Press.

Kong, Lily, and Brenda S. A. Yeoh. 2003. *The Politics of Landscapes in Singapore: Constructions of "nation."* Syracuse, NY: Syracuse University Press.

Labinger, Jay A., and Harry Collins, eds. 2001. *The One Culture? A Conversation about Science.* Chicago: University of Chicago Press.

Lambeck, Kurt, Hélène Rouby, Anthony Purcell, Yiying Sun, and Malcolm Sambridge. 2014. "Sea Level and Global Ice Volumes from the Last Glacial Maximum to the Holocene." *Proceedings of the National Academy of Sciences* 111 (43): 15296–303.

Landes, David S. 1999. *The Wealth and Poverty of Nations: Why Some Are So Rich and Some So Poor.* New York: Norton.

———. 2006. "Why Europe and the West? Why Not China?" *Journal of Economic Perspectives* 20 (2): 3–22.

Langhans, Simone D., Judit Lienert, Nele Schuwirth, and Peter Reichert. 2013. "How to Make River Assessments Comparable: A Demonstration for Hydromorphology." *Ecological Indicators* 32 (September): 264–75.

Large, A. R. G., and D. J. Gilvear. 2015. "Using Google Earth, a Virtual-Globe Imaging Platform, for Ecosystem Services-Based River Assessment." *River Research and Applications* 31 (4): 406–21.

Lassaletta, Luis, Gilles Billen, Bruna Grizzetti, Josette Garnier, Allison M. Leach, and James N. Galloway. 2014. "Food and Feed Trade as a Driver in the Global Nitrogen Cycle: 50-Year Trends." *Biogeochemistry* 118 (1–3): 225–41.

Latour, Bruno. 1988. *Science in Action: How to Follow Scientists and Engineers through Society.* Reprint ed. Cambridge, MA: Harvard University Press.

———. 2005. *Reassembling the Social: An Introduction to Actor-Network-Theory.* Oxford: Oxford University Press.

———. 2013. "Facing Gaia: A New Inquiry into Natural Religion." Lecture presented at the Gifford Lectures 2013, Edinburgh, U.K., February 18. http://www.giffordlectures.org/search/site/2013.

Latour, Bruno, and Steve Woolgar. 1986. *Laboratory Life: The Construction of Scientific Facts.* Reprint ed. Princeton, NJ: Princeton University Press.

Lautze, Sue, Daniel Maxwell, Stephen Devereux, et al. 2007. "Why Do Famines Persist in the Horn of Africa? Ethiopia, 1999–2003." In *The New Famines: Why Famines Persist in an Era of Globalization,* edited by Stephen Devereaux, 222–44. Routledge Studies in Development Economics. London: Routledge.

Leavy, Patricia. 2012. *Essentials of Transdisciplinary Research: Using Problem-Centered Methodologies.* San Francisco: Left Coast Press.

Le Cloarec, M.-F., P. H. Bonte, L. Lestel, I. Lefèvre, and S. Ayrault. 2011. "Sedimentary Record of Metal Contamination in the Seine River during the Last Century." *Physics and Chemistry of the Earth, Parts A/B/C* 36 (12): 515–29.

Lee, E. T. 2015. "The Search for NEWater: The Singapore Water Story." In *50 Years of Environment: Singapore's Journey towards Environmental Sustainability,* edited by Tan Yong Soon, 63–72. Hackensack, NJ: World Scientific Publishing Co.

Lee, Kai N. 1999. "Appraising Adaptive Management." *Ecology and Society* 3 (2).

Lee, L. Y., and C. N. Ong. 2015. "Frontier Research in Environment and Water: Integrated Research Approach for Sustainable Solutions." In *50 Years of Environment: Singapore's Journey Towards Environmental Sustainability,* edited by Tan Yong Soon, 85–125. Hackensack, NJ: World Scientific Publishing Co.

Lee, P. O. 2003. "The Water Issue between Singapore and Malaysia: No Solution in Sight?" http://agris.fao.org/agris-search/search.do?recordID=GB2013200036.

Leidreiter, Anna. 2010. "Community Participation in Common Natural Resource Management in the Lake Tana Watershed, Ethiopia." MSc. thesis, Universiteit van Amsterdam. http://131.220.109.9/fileadmin/webfiles/downloads/research_docu/Leidreiter_2010_Community_participation.pdf.

Leopold, Aldo. 1968. *A Sand County Almanac and Sketches Here and There.* Oxford: Oxford University Press.

Lepore, Jill. 2001. "Historians Who Love Too Much: Reflections on Microhistory and Biography." *Journal of American History* 88 (1): 129–44.

Lestel, Laurence. 2012. "Non-Ferrous Metals (Pb, Cu, Zn) Needs and City Development: The Paris Example (1815–2009)." *Regional Environmental Change* 12 (2): 311–23.

Lestel, Laurence, and Cau Carré, eds. 2017. *Les rivières urbaines et leur pollution.* Indisciplines. Paris: Quae Editions.

Lestel, Laurence, Simona Georgescu, Pierre Alexandre, Julie Davodet, Joséphine Roulliard, Alexandre Galibert, and Aurélien Baro. 2015. "Cartographie historique du bassin de la Seine." Rapport de fin de phase VI du PIREN Seine. http://www.metis.upmc.fr/piren/?q=webfm_send/1495.

Lestel, Laurence, Michel Meybeck, and Daniel Thevenot. 2007. "Metal Contamination Budget at the River Basin Scale: An Original Flux-Flow Analysis (F2A) for the Seine River." *Hydrology and Earth System Sciences Discussions* 11 (6): 1771–81.

Lewis, Simon L., and Mark A. Maslin. 2015. "Defining the Anthropocene." *Nature* 519 (7542): 171–80.

Lim, M. C. 1997. "Drainage Planning and Control in the Urban Environment." *Environmental Monitoring and Assessment* 44: 44, 183, 189, 194–97.

Lovelock, James. 1974. "Atmospheric Homeostasis by and for the Biosphere: The Gaia Hypothesis." *Tellus* 26: 1–9.

———. 1983. "Gaia as Seen through the Atmosphere." In *Biomineralization and Biological Metal Accumulation,* edited by Peter Westbroek and E. W. de Jong, 15–25. Dordrecht: D. Reidel.

———. [1987] 2000. *Gaia: A New Look at Life on Earth.* 2nd ed. Oxford: Oxford University Press.

———. 2006. *The Revenge of Gaia: Earth's Climate Crisis & the Fate of Humanity.* New York: Penguin.

Lubinski, K. 2010. "The Concept of River Ecosystem Health and Its Relationship to Management." Workshop Proceedings presented at the US-Indo Bilateral Workshop on Sedimentation, Erosion, Flooding and Ecological Health of Rivers, Indiana Statistical Institute, Kolkata, India.

Ludi, Eva. 2009. "Climate Change, Water and Food Security." Background Note. Overseas Development Institute. http://dspace.africaportal.org/jspui/bitstream/123456789/24103/1/Climate%20change%20-%20water%20and%20food%20security.pdf?1.

Lyell, Charles. 1832. *Principles of Geology, Being an Attempt to Explain the Former Changes of the Earth's Surface, by Reference to Causes Now in Operation.* Vol. 2. London: John Murray.

Machlis, Gary E., Jo Ellen Force, and William R. Burch Jr. 1997. "The Human Ecosystem Part I: The Human Ecosystem as an Organizing Concept in Ecosystem Management." *Society & Natural Resources* 10 (4): 347–67.

MacKenzie, John M. 1997. *The Empire of Nature: Hunting, Conservation and British Imperialism.* Manchester: Manchester University Press.

Malay Heritage Centre. 2013. Permanent Gallery Exhibit. Singapore.

Malim, Tim. 2005. *Stonea and the Roman Fens.* Stroud, U.K.: Tempus Publications.

Malm, Andreas, and Alf Hornborg. 2014. "The Geology of Mankind? A Critique of the Anthropocene Narrative." *Anthropocene Review* 1 (1): 62–69.

Margulis, Lynn. 2008. *Symbiotic Planet: A New Look at Evolution*. New York: Basic Books.

Marsh, George Perkins. 1864. *Man and Nature; Or, Physical Geography as Modified by Human Action*. New York: Charles Scribner.

Mattor, Katherine, Michelle Betsill, Ch'aska Huayhuaca, Heidi Huber-Stearns, Faith Sternleib, Patrick Bixler, Antony Cheng, and Matthew Luizza. 2013. "Transdisciplinary Research on Environmental Governance: A View from the Trenches." 29. Earth System Governance Working Paper.

Mauch, Christof, and Thomas Zeller, eds. 2008. *Rivers in History: Perspectives on Waterways in Europe and North America*. Pittsburgh, PA: University of Pittsburgh Press.

McCormick, John. 1991. *Reclaiming Paradise: The Global Environmental Movement*. Bloomington: Indiana University Press.

Meadows, Donella, Dennis Meadows, Jørgen Randers, and William Behrens III. 1972. *Limits to Growth*. New York: Club of Rome.

Mekonnen, Mesfin M., and Arjen Y. Hoekstra. 2011. "National Water Footprint Accounts: The Green, Blue and Grey Water Footprint of Production and Consumption." 50. Delft: UNESCO-IHE Institute for Water Education. http://www.unesco-ihe.org/Value-of-Water-Research-Report-Series/Research-Papers.

Merritts, Dorothy, Robert Walter, Michael Rahnis, Jeff Hartranft, Scott Cox, Allen Gellis, Noel Potter, et al. 2011. "Anthropocene Streams and Base-Level Controls from Historic Dams in the Unglaciated Mid-Atlantic Region, USA." *Philosophical Transactions: Mathematical, Physical and Engineering Sciences* 369 (1938): 976–1009.

Meybeck, M. 1986. "Composition chimique des ruisseaux non pollués de France." *Sciences Geologique (Bulletin)* 39 (1): 3–77.

———. 1998. "Man and River Interface: Multiple Impacts on Water and Particulates Chemistry Illustrated in the Seine River Basin." In *Oceans, Rivers and Lakes: Energy and Substance Transfers at Interfaces,* 131: 1–20. Developments in Hydrobiology. New York: Springer.

———. 2001. "River Basin under Anthropocene Conditions." In *Science and Integrated Basin Management,* 275–94. Dahlem Workshop Series. New York: Wiley.

———. 2002. "Riverine Quality at the Anthropocene: Propositions for Global Space and Time Analysis, Illustrated by the Seine River." *Aquatic Sciences* 64 (4): 376–93.

———. 2003. "Global Analysis of River Systems: From Earth System Controls to Anthropocene Syndromes." *Philosophical Transactions of the Royal Society of London B: Biological Sciences* 358 (1440): 1935–55.

Meybeck, Michel, and Richard Helmer. 1989. "The Quality of Rivers: From Pristine Stage to Global Pollution." *Global and Planetary Change* 1 (4): 283–309.

Meybeck, Michel, Laurence Lestel, Philippe Bonté, Régis Moilleron, Jean Louis Colin, Olivier Rousselot, Daniel Hervé, Claire De Pontevès, Cécile Grosbois, and Daniel R. Thévenot. 2007. "Historical Perspective of Heavy Metals Contamination (Cd, Cr, Cu, Hg, Pb, Zn) in the Seine River Basin (France) Following a DPSIR Approach (1950–2005)." *Science of the Total Environment* 375 (1): 204–31.

Meybeck, M., L. Lestel, C. Carré, G. Bouleau, J. Garnier, and J.M. Mouchel. 2016. "Trajectories of River Chemical Quality Issues over the Longue Durée: The Seine River (1900s–2010)." *Environmental Science and Pollution Research*. doi:10.1007/s11356-016-7124-0.

Meybeck, Michel, Ghislain de Marsily, and Éliane Fustec, eds. 1998. *La Seine en son bassin: Fonctionnement écologique d'un système fluvial anthropisé.* Philadelphia, PA: Elsevier.

Meybeck, Michel, and Charles Vörösmarty. 2005. "Fluvial Filtering of Land-to-Ocean Fluxes: From Natural Holocene Variations to Anthropocene." *Comptes Rendus Geoscience* 337 (1–2): 107–23.

Mill, John Stuart. 1878. *Principles of Political Economy: With Some of Their Applications to Social Philosophy.* London: Longmans, Green, Reader, and Dyer.

Millennium Ecosystem Assessment (MEA). 2005. *Ecosystems and Human Well-Being.* Vol. 200. Washington, DC: Island Press. http://www.who.int/entity/globalchange/ecosystems/ecosys.pdf.

Ministère de l'écologie, du développement durable et de l'énergie. 2012. "Évaluation de l'état des eaux de surface continentales (cours d'eau, canaux, plans d'eau)." Guide technique.

Ministry of Water Resources (MoWR). 2001. "Ethiopian Water Sector Strategy." Ethiopian Ministry of Water Resources, Addis Ababa. www.mowr.gov.et%2Fattachmentfiles%2FDownloads%2FWater%2520policy.doc&rct=j&q=promoting the principles of integrated water resources management ministry water resources ethiopia 2001&ei=SYNfTt2dFYnz-gaBzun9AQ&usg=AFQjCNHWyf_GoJGv-FYElawWXhtsUJy5Q&sig2=FH2bjMOELoMMYb-Otzywsg&cad=rja.

Ministry of Water Resources and National Meteorological Agency (MoWR and NMA). 2007. "National Adaptation Programme of Action of Ethiopia." Addis Ababa: Ethiopian Ministry of Water Resources; National Meteorological Agency.

Mitchell, Timothy. 2002. *Rule of Experts: Egypt, Techno-Politics, Modernity.* Berkeley: University of California Press.

MMD (Mott MacDonald). 2005. "Koga Irrigation Project: Water User Associations." 12. Working Paper. Ministry of Water Resources, Addis Ababa.

Moges, Semu, Helmut Kloos, Stuart McFeeters, and Worku Legesse. 2010. "The Water Resources of Ethiopia and Large-Scale Hydropower and Irrigation Development." In *Water Resources Management in Ethiopia: Implications for the Nile Basin,* edited by Helmut Kloos and Worku Legesse, 63–101. Amherst, NY: Cambria Press.

Molle, François. 2004. *Evolution of Irrigation in South and Southeast Asia.* Colombo, Sri Lanka: IWMI.

Mollinga, Peter P. 2010. "Boundary Work and the Complexity of Natural Resources Management." *Crop Science* 50 (Supplement 1): S-1–S-9.

Monastersky, Richard. 2014. "Biodiversity: Life—a Status Report." *Nature News* 516 (7530): 158.

Moore, Jason W. 2003. "The Modern World-System as Environmental History? Ecology and the Rise of Capitalism." *Theory and Society* 32 (3): 307–77.

Moore, T. n.d. *Philosophy for Dummies.* New York: Wiley.

Morton, Timothy. 2013. *Hyperobjects: Philosophy and Ecology after the End of the World.* Minneapolis: University of Minnesota Press.

Mosley, Stephen. 2006. "Common Ground: Integrating Social and Environmental History." *Journal of Social History* 39 (3): 915–33.

Mouchel, J. M., and G. Billen, eds. 2008–15. *The Seine River Basin.* [Collection of booklets.] http://metis.upmc.fr/piren/.

Mrazek, David, dir. 2014. *From Billions to None: The Passenger Pigeon's Flight to Extinction.* Documentary.

Myers, Norman, et al. 1979. *The Sinking Ark: A New Look at the Problem of Disappearing Species.* Oxford: Pergamon Press.

Naidoo, R., A. Balmford, R. Costanza, B. Fisher, R. E. Green, B. Lehner, T. R. Malcolm, and T. H. Ricketts. 2008. "Global Mapping of Ecosystem Services and Conservation Priorities." *Proceedings of the National Academy of Sciences* 105 (28): 9495–9500.

Naiman, Robert J. 1983. "The Annual Pattern and Spatial Distribution of Aquatic Oxygen Metabolism in Boreal Forest Watersheds." *Ecological Monographs* 53 (1): 73–94.

Narrative of the Great Flood in the Rivers Tyne, Tease [sic], Wear, &c. 1772. Newcastle.

Nelson, Erik, Guillermo Mendoza, James Regetz, Stephen Polasky, Heather Tallis, Drichard Cameron, Kai M. A. Chan, et al. 2009. "Modeling Multiple Ecosystem Services, Biodiversity Conservation, Commodity Production, and Tradeoffs at Landscape Scales." *Frontiers in Ecology and the Environment* 7 (1): 4–11.

Neumann, Rod. 2005. *Making Political Ecology.* London: Routledge.

Newcastle Common Council. 1772. "Unpublished Minutes." Tyne and Wear Records Office, Newcastle-upon-Tyne.

Newson, Malcolm D., and Andrew R. G. Large. 2006. "'Natural' Rivers, 'Hydromorphological Quality' and River Restoration: A Challenging New Agenda for Applied Fluvial Geomorphology." *Earth Surface Processes and Landforms* 31 (13): 1606–24.

Newson, Malcolm D., John Pitlick, and David A. Sear. 2002. "Running Water: Fluvial Geomorphology and River Restoration." In *Handbook of Ecological Restoration,* vol. 1, *Principles,* edited by Martin R. Perrow and Anthony J. Davy, 1:133–52. Cambridge: Cambridge University Press.

Nicolescu, Basarab. 2014. "Methodology of Transdisciplinarity." *World Futures* 70 (3–4): 186–99.

Northumberland County Council. 2010. "Level 1 Strategic Flood Risk Assessment; Appendix A: Historical Flood Events." http://www.northumberland.gov.uk/WAM Documents/1F474BCD-B02E-46FD-BFB0-2FAA7E12A41F_1_0.pdf?nccredirect=1.

Oglesby, Ray, Clarence Carlson, and James McCann, eds. 1972. *River Ecology and Man.* New York: Academic Press.

Olsen, Randy. 2002. "Shifting Baselines: The Truth about Ocean Decline." November 17. http://www.shiftingbaselines.org/op_ed/.

Ong, A. 2004. "Island Nations." In *Patterned Ground: Entanglements of Nature and Culture,* edited by Stephan Harrison, Steve Pile, and Nigel Thrift, 270–72. London: Reaktion Books.

Osterhammel, Jürgen. 2014. *The Transformation of the World: A Global History of the Nineteenth Century.* Translated by Patrick Camiller. Princeton, NJ: Princeton University Press.

Ostrom, Elinor. 1992. "Crafting Institutions for Self-Governing Irrigation Systems." *River Research and Application* 8 (3): 307–9.

———. 1993. "Design Principles in Long-Enduring Irrigation Institutions." *Water Resources Research* 29 (7): 1907–12.

———. 2009. "A General Framework for Analyzing Sustainability of Social-Ecological Systems." *Science* 325 (5939): 419–22.

Ottinger, Gwen, and Benjamin R. Cohen, eds. 2011. *Technoscience and Environmental Justice: Expert Cultures in a Grassroots Movement*. Cambridge, MA: MIT Press.

Oudin, Louis-Charles, and Danielle Maupas. 1999. "Système d'évaluation de la qualité de l'eau des cours d'eau, rapport de présentation SEQ-Eau (version 1)." Étude Inter Agences.

Palsson, Gisli, Bronislaw Szerszynski, Sverker Sörlin, John Marks, Bernard Avril, Carole Crumley, Heide Hackmann, et al. 2013. "Reconceptualizing the 'Anthropos' in the Anthropocene: Integrating the Social Sciences and Humanities in Global Environmental Change Research." *Environmental Science & Policy* 28: 3–13.

Papers Relating to Land Tenures and Revenue Settlement in Oude. 1865. Calcutta: O. T. Cutter, Military Orphan Press.

Papworth, S. K., J. Rist, L. Coad, and E. J. Milner-Gulland. 2009. "Evidence for Shifting Baseline Syndrome in Conservation." *Conservation Letters* 2 (2): 93–100.

Parthasarathi, Prasannan. 2011. *Why Europe Grew Rich and Asia Did Not: Global Economic Divergence, 1600–1850*. 1st ed. Cambridge: Cambridge University Press.

Passy, Paul, Josette Garnier, Gilles Billen, Corinne Fesneau, and Julien Tournebize. 2012. "Restoration of Ponds in Rural Landscapes: Modelling the Effect on Nitrate Contamination of Surface Water (the Seine River Basin, France)." *Science of the Total Environment* 430: 280–90.

Pastore, Christopher L., Mark B. Green, Daniel J. Bain, Andrea Muñoz-Hernandez, Charles J. Vörösmarty, Jennifer Arrigo, Sara Brandt, et al. 2010. "Tapping Environmental History to Recreate America's Colonial Hydrology." *Environmental Science & Technology* 44 (23): 8798–8803.

"Paul J. Crutzen—Biographical." 1995. Nobelprize.org. http://www.nobelprize.org/nobel_prizes/chemistry/laureates/1995/crutzen-bio.html.

Pauly, Daniel. 1995. "Anecdotes and the Shifting Baseline Syndrome of Fisheries." *Trends in Ecology & Evolution* 10 (10): 430.

Pausewang, S. 2002. "No Environmental Protection without Local Democracy? Why Peasants Distrust Their Agricultural Advisers." In *Ethiopia: The Challenge of Democracy from Below*, edited by Bahru Zewde and Siegfried Pausewang, 87–102. Stockholm: Elanders Gotab.

Pearce, Fred. 2008. *With Speed and Violence: Why Scientists Fear Tipping Points in Climate Change*. 1st ed. Boston: Beacon Press.

Peet, Richard, and Michael Watts. 1993. "Introduction: Development Theory and Environment in an Age of Market Triumphalism." *Economic Geography* 69 (3): 227–53.

———, eds. 2004. *Liberation Ecologies*. 2nd ed. London: Routledge.

Piaget, Jean. 1974. "L'épistémologie des relations interdisciplinaires." *Internationales Jahrbuch für Interdisziplinäre Forschung* 1: 154–72.

Price, S. J., J. R. Ford, A. H. Cooper, and C. Neal. 2011. "Humans as Major Geological and Geomorphological Agents in the Anthropocene: The Significance of Artificial Ground in Great Britain." *Philosophical Transactions of the Royal Society A: Mathematical, Physical and Engineering Sciences* 369 (1938): 1056–84.

Pryor, Francis. 2005. *Flag Fen: Life and Death of a Prehistoric Landscape*. Stroud, U.K.: Tempus Publications.

Public Utilities Board (PUB). 2012. "Innovation in Water Singapore." Singapore.

Quash, Ben. 2005. *Theology and the Drama of History.* Cambridge: Cambridge University Press.

Raffles, Hugh. 2002. *In Amazonia: A Natural History.* Princeton, NJ: Princeton University Press.

Randi Korn & Associates, Inc. (RK&A). 2012. "FLOW: Can You See the River?" Project evaluation report. Unpublished manuscript. Indianapolis Museum of Art, Indianapolis, IN.

Rinne, John N., ed. 2005. *Historical Changes in Large River Fish Assemblages of the Americas (American Fisheries Society Symposium).* Bethesda, MD: American Fisheries Society.

Robin, Libby, and Will Steffen. 2007. "History for the Anthropocene." *History Compass* 5 (5): 1694–1719.

Rockström, Johan, W. Steffen, Kevin Noone, Åsa Persson, F. Stuart Chapin, Eric Lambin, Timothy Lenton, Marten Scheffer, Carl Folke, Hans Schellnhuber, Björn Nykvist, Cynthia de Wit, Terry Hughes, Sander van der Leeuw, Henning Rodhe, Sverker Sörlin, et al. 2009a. "Planetary Boundaries: Exploring the Safe Operating Space for Humanity." *Ecology and Society,* January.

———. 2009b. "A Safe Operating Space for Humanity." *Nature* 461 (7263): 472–75.

Romero, Estela, Romain Le Gendre, Josette Garnier, Gilles Billen, Cédric Fisson, Marie Silvestre, and Philippe Riou. 2016. "Long-Term Water Quality in the Lower Seine: Lessons Learned over 4 Decades of Monitoring." *Environmental Science & Policy* 58: 141–54.

Roosevelt, Theodore. 1910. *African Game Trails: An Account of the African Wanderings of an American Hunter-Naturalist.* New York: Syndicate Publishing Co.

Rotman, Michael. n.d. "Cuyahoga River Fire." *Cleveland Historical.* https://clevelandhistorical.org/items/show/63#.Ut9BYbROnZ4.

Rubenstein, Michael, and Robert John Russell. 2010. *Public Works.* Notre Dame, IN: University of Notre Dame Press.

Ruddiman, William F. 2003. "The Anthropogenic Greenhouse Era Began Thousands of Years Ago." *Climatic Change* 61 (3): 261–93.

———. 2007. "The Early Anthropogenic Hypothesis: Challenges and Responses." *Reviews of Geophysics* 45 (4): RG4001.

———. 2013. "The Anthropocene." *Annual Review of Earth and Planetary Sciences* 41 (1): 45–68.

"The Salmon Fisheries Conference [Horticultural Gardens, South Kensington, 7th June 1867]." 1867. *Journal of Agriculture,* 3rd, December.

Sassen, Saskia. 2006. *Territory, Authority, Rights: From Medieval to Global Assemblages.* Vol. 4. Cambridge: Cambridge University Press.

Scalise, Carmen, and Kevin Fitzpatrick. 2012. "Chicago Deep Tunnel Design and Construction." In *Structures Congress 2012,* 1485–95. Reston, VA: American Society of Civil Engineers.

Scarpino, Philip V. 1997. "Large Floodplain Rivers as Human Artifacts: A Historical Perspective on Ecological Integrity." DTIC Document. http://oai.dtic.mil/oai/oai?verb=getRecord&metadataPrefix=html&identifier=ADA337599.

———. 2010. "Addressing Cross-Border Pollution of the Great Lakes." In *Transnationalism: Canada–United States History into the Twenty-First Century,* edited by Michael Behiels and Reginald C. Stuart. Montreal: McGill-Queen's University Press.

———. 2014. "Rivers of the Anthropocene." Paper presented at the Rivers of the Anthropocene Conference, Indianapolis, IN.

Schaich, Harald, Claudia Bieling, and Tobias Plieninger. 2010. "Linking Ecosystem Services with Cultural Landscape Research." *GAIA—Ecological Perspectives for Science and Society* 19 (4): 269–77.

Schlereth, Thomas J. 1989. "History Museums and Material Culture." In *History Museums in the United States: A Critical Assessment,* edited by Warren Leon and Roy Rosenzweig, 294–320. Urbana: University of Illinois Press.

Schmidt, Morgan J., Anne Rapp Py-Daniel, Claide de Paula Moraes, Raoni B. M. Valle, Caroline F. Caromano, Wenceslau G. Texeira, Carlos A. Barbosa, et al. 2014. "Dark Earths and the Human Built Landscape in Amazonia: A Widespread Pattern of Anthrosol Formation." *Journal of Archaeological Science* 42 (February): 152–65.

Schwägerl, Christian. 2014. *The Anthropocene: The Human Era and How It Shapes Our Planet.* Santa Fe, NM: Synergetic Press.

Scott, Peter. 2011. "Right Out of Time? Politics and Nature in a Postnatural Condition." In *Religion and Ecology in the Public Sphere,* edited by Celia Deane-Drummond and Heinrich Bedford-Strohm, 57–76. New York: Continuum/T&T Clark International.

Seitzinger, S. P., J. A. Harrison, Egon Dumont, Arthur H. W. Beusen, and A. F. Bouwman. 2005. "Sources and Delivery of Carbon, Nitrogen, and Phosphorus to the Coastal Zone: An Overview of Global Nutrient Export from Watersheds (NEWS) Models and Their Application." *Global Biogeochemical Cycles* 19 (4): GB4S01. doi:10.1029/2005GB002606.

Servais, Pierre, Gilles Billen, A. Goncalves, and Tamara Garcia-Armisen. 2007. "Modelling Microbiological Water Quality in the Seine River Drainage Network: Past, Present and Future Situations." *Hydrology and Earth System Sciences Discussions* 11 (5): 1581–92.

Shapin, Steven. 1995. *A Social History of Truth: Civility and Science in Seventeenth-Century England.* Chicago: University of Chicago Press.

———. 1996. *The Scientific Revolution.* Chicago: University of Chicago Press.

Shapin, Steven, and Simon Schaffer. 1986. *Leviathan and the Air-Pump: Hobbes, Boyle, and the Experimental Life.* Princeton, NJ: Princeton University Press.

Shenkut, Mammo Kebbede. 2006. "Ethiopia: Where and Who Are the World's Illiterates?" EFA Global Monitoring Report 2006, Literacy for Life. http://datatopics.worldbank.org/hnp/files/edstats/ethgmrpro05.pdf.

Shennan, Ian, and Ben Horton. 2002. "Holocene Land- and Sea-Level Changes in Great Britain." *Journal of Quaternary Science* 17 (5–6): 511–26.

Shivakoti, G. P., and Elinor Ostrom. 2016. "Improving Irrigation Governance and Management in Nepal," April. https://vtechworks.lib.vt.edu/handle/10919/65944.

Skalak, Katherine J., Adam J. Benthem, Edward R. Schenk, Cliff R. Hupp, Joel M. Galloway, Rochelle A. Nustad, and Gregg J. Wiche. 2013. "Large Dams and Alluvial Rivers in the Anthropocene: The Impacts of the Garrison and Oahe Dams on the Upper Missouri River." *Anthropocene*, Geomorphology of the Anthropocene: Understanding the Surficial Legacy of Past and Present Human Activities, 2 (October): 51–64.

Skertchly, S. B. J. 1877. *The Geology of the Fenland: Memoirs of the Geological Survey, England and Wales.* London: HMSO.

Smith, Dinah M., Jan A. Zalasiewicz, Mark Williams, Ian P. Wilkinson, Martin Redding, and Crane Begg. 2010. "Holocene Drainage Systems of the English Fenland: Roddons and Their Environmental Significance." *Proceedings of the Geologists' Association* 121 (3): 256–69.

Smith, Dinah M., Jan A. Zalasiewicz, Mark Williams, Ian P. Wilkinson, James J. Scarborough, Mark Knight, Carl Sayer, Martin Redding, and Steven G. Moreton. 2012. "The Anatomy of a Fenland Roddon: Sedimentation and Environmental Change in a Lowland Holocene Tidal Creek Environment." *Proceedings of the Yorkshire Geological Society* 59 (2): 145–59.

Smith, Bruce D., and Melinda A. Zeder. 2013. "The Onset of the Anthropocene." *Anthropocene,* When Humans Dominated the Earth: Archeological Perspectives on the Anthropocene, 4 (December): 8–13.

Society of Antiquaries of Newcastle (SANT). 1771. "River Tyne Flood Papers." Hancock Museum, Newcastle Upon Tyne.

Sokal, Alan, and Jean Bricmont. 1999. *Fashionable Nonsense: Postmodern Intellectuals' Abuse of Science.* New York: Picador.

Solzman, David M. 1998. *The Chicago River: An Illustrated History and Guide to the River and Its Waterways.* Chicago: University of Chicago Press.

Sörlin, Sverker. 2012. "Environmental Humanities: Why Should Biologists Interested in the Environment Take the Humanities Seriously?" *BioScience* 62 (9): 788–89.

Sponseller, Ryan A., James B. Heffernan, and Stuart G. Fisher. 2013. "On the Multiple Ecological Roles of Water in River Networks." *Ecosphere* 4 (2): 1–14.

Stanford, Jack A., and Geoffrey C. Poole. 1996. "A Protocol for Ecosystem Management." *Ecological Applications* 6 (3): 741–44.

Starkey, Eleanor and Geoff Parkin. 2015. "Community Involvement in UK Catchment Management." *Review of Current Knowledge.* Marlow, Buckinghamshire: Foundation for Water Research. http://www.fwr.org/Catchment/frr0021.pdf.

Starkey, Eleanor, Geoff Parkin, Stephen Birkinshaw, Andy Large, Paul Quinn and Ceri Gibson. 2017. "Demonstrating the Value of Community-Based ('Citizen Science') Observations for Catchment Modelling and Characterisation." *Journal of Hydrology* 548: 801–817

Steffen, Will, Wendy Broadgate, Lisa Deutsch, Owen Gaffney, and Cornelia Ludwig. 2015. "The Trajectory of the Anthropocene: The Great Acceleration." *Anthropocene Review* 2 (1): 81–98.

Steffen, Will, Paul Crutzen, and John McNeill. 2007. "The Anthropocene: Are Humans Now Overwhelming the Great Forces of Nature." *AMBIO: A Journal of the Human Environment* 36 (8): 614–21.

Steffen, Will, Paul Crutzen, James McNeill, and Kathy A. Hibbard. 2008. "Stages of the Anthropocene: Assessing the Human Impact on the Earth System." *AGU Fall Meeting Abstracts,* December. http://adsabs.harvard.edu/abs/2008AGUFMGC22B.01S.

Steffen, Will, Jacques Grinevald, Paul Crutzen, and John McNeill. 2011. "The Anthropocene: Conceptual and Historical Perspectives." *Philosophical Transactions of the Royal Society A: Mathematical, Physical and Engineering Sciences* 369 (1938): 842–67.

Steffen, Will, Åsa Persson, Lisa Deutsch, Jan Zalasiewicz, Mark Williams, K. Richardson, C. Crumley, et al. 2011. "The Anthropocene: From Global Change to Planetary Stewardship." *AMBIO: A Journal of the Human Environment,* 1–23.

Steffen, Will, Angelina Sanderson, Peter Tyson, Jill Jäger, P. A. Matson, F. Oldfield, K. Richardson, H.-J. Schellnhuber, B. L. Turner II, and R. J. Wasson. 2004. *Global Change and the Earth System: A Planet under Pressure*. Berlin: Springer.

Steinhart, Edward I. 2006. *Black Poachers, White Hunters: A Social History of Hunting in Colonial Kenya*. Athens: Ohio University Press.

Stocker, Thomas F., D. Qin, G. K. Plattner, M. Tignor, S. K. Allen, J. Boschung, A. Nauels, Y. Xia, V. Bex, and P. M. Midgley. 2013. "Climate Change 2013: The Physical Science Basis. Intergovernmental Panel on Climate Change, Working Group I Contribution to the IPCC Fifth Assessment Report (AR5)." Intergovernmental Panel on Climate Change, Working Group I Contribution to the IPCC Fifth Assessment Report (AR5). New York: IPCC.

Strickland, Henry Edwin. 1848. "History and External Characters of the Dodo, Solitaire, and Other Extinct Brevipennate Birds of Mauritius, Rodriguez, and Bourbon." In *The Dodo and Its Kindred: Or, The History, Affinities, and Osteology of the Dodo, Solitaire, and Other Extinct Birds of the Islands Mauritius, Rodriguez, and Bourbon*, edited by Alexander Gordon Melville and Henry Edwin Strickland. London: Reeve, Benham, and Reeve.

Stroup, George W. 1984. *The Promise of Narrative Theology*. London: SCM Press.

Subcommission on Quaternary Stratigraphy, International Commission on Stratigraphy. 2015. "Working Group on the 'Anthropocene.'" Accessed January 21. http://quaternary.stratigraphy.org/workinggroups/anthropocene/.

Swain, Ashok. 2002. "The Nile River Basin Initiative: Too Many Cooks, Too Little Broth." *SAIS Review* 22 (2): 293–308.

Swimme, Brian, and Thomas Berry. 1992. *The Universe Story*. San Francisco, CA: HarperCollins.

Syvitski, James. 2012. "Anthropocene: An Epoch of Our Making." *Global Change* 78 (March): 12–15.

———. 2014. "Rivers of the Anthropocene." Paper presented at the Rivers of the Anthropocene Conference, IUPUI, January. https://www.youtube.com/watch?v=CnC7mAGuJ1c.

———. 2016a. "The Anthropocene—From Concept, to Geological Epoch, to 21st Century Science and Public Discourse." Lecture, Indiana University, Bloomington, April 14.

———. 2016b. "Welcome to the Anthropocene: A Brief History of How Humans Are Shaping the Planet." Presented at the Jack Jeffries Public Lecture, Newcastle University, July 11.

Syvitski, James P. M., and Albert Kettner. 2011. "Sediment Flux and the Anthropocene." *Philosophical Transactions of the Royal Society of London A: Mathematical, Physical and Engineering Sciences* 369 (1938): 957–75.

Syvitski, James P. M., Charles J. Vörösmarty, Albert J. Kettner, and Pamela Green. 2005. "Impact of Humans on the Flux of Terrestrial Sediment to the Global Coastal Ocean." *Science* 308 (5720): 376–80.

Sze, Julie, and Jonathan K. London. 2008. "Environmental Justice at the Crossroads." *Sociology Compass* 2 (4): 1331–54.

Szerszynski, Bronislaw. 2012. "The End of the End of Nature: The Anthropocene and the Fate of the Human." *Oxford Literary Review* 34 (2): 165–84.

Tabuchi, Jean-Pierre, Lionel Benard, Béatrice Blanchet, Fabien Esculier, Jean-Marie Mouchel, Michel Poulin, and Olivier Saison. 2013. "Gestion en temps réel du système d'assainissement de la région parisienne en fonction de la qualité de la Seine." *NOVATECH 2013*. http://documents.irevues.inist.fr/handle/2042/51360.

Tales, Evelyne, Jérôme Belliard, Guillaume Gorges, and Céline Le Pichon, eds. 2009. *Le peuplement de poissons du bassin de la Seine.* Vol. 4. Programme Piren-Seine. Eau Seine Normandie. http://www.metis.upmc.fr/piren/?q=book/977.

Tan, A. 2016. "Singapore as a Global Hub Port and International Maritime Centre." In *Singapore 2065: Leading Insights on Economy and Environment from 50 Singapore Icons and Beyond.* Hackensack, NJ: World Scientific Publishing Co.

Task Force. 2012. "Expert Panel Report." Singapore.

Taylor, N.H. 1934. "The Alleviation of Flooding in Singapore." Malayan Association of Institute of Civil Engineers. Mimeograph.

Theriot, Christopher and Tzoumis, Kelly. 2007. "Bankside Chicago." In *Rivertown: Rethinking Urban Rivers,* edited by Paul S. Kibel, 67–83. Cambridge, MA: MIT Press.

Thévenot, Daniel R., Régis Moilleron, Laurence Lestel, Marie-Christine Gromaire, Vincent Rocher, Philippe Cambier, Philippe Bonté, Jean-Louis Colin, Claire De Pontevès, and Michel Meybeck. 2007. "Critical Budget of Metal Sources and Pathways in the Seine River Basin (1994–2003) for Cd, Cr, Cu, Hg, Ni, Pb and Zn." *Science of the Total Environment* 375 (1): 180–203.

Thomas, Julia Adeney. 2014. "History and Biology in the Anthropocene: Problems of Scale, Problems of Value." *American Historical Review* 119 (5): 1587–1607.

Thorp, James H., Martin C. Thoms, and Michael D. Delong. 2006. "The Riverine Ecosystem Synthesis: Biocomplexity in River Networks across Space and Time." *River Research and Applications* 22 (2): 123–47.

———. 2008. *The Riverine Ecosystem Synthesis: Toward Conceptual Cohesiveness in River Science.* London: Academic Press.

"Time for the Social Sciences." 2014. *Nature* 517 (7532): 5.

Timmons Roberts, J. 2007. "Globalizing Environmental Justice." In *Environmental Justice and Environmentalism: The Social Justice Challenge to the Environmental Movement,* edited by Ronald D. Sandler and Phaedra C. Pezzullo, 285–308. Cambridge, MA: MIT Press.

Tortajada, Cecilia. 2006. "Water Management in Singapore." *International Journal of Water Resources Development* 22 (2): 227–40.

Turner, B.L., William Clark, Robert Kates, John Richards, Jessica Mathews, and William Meyer, eds. 1990. *The Earth as Transformed by Human Action: Global and Regional Changes in the Biosphere over the Past 300 Years.* Cambridge: Cambridge University Press.

Tzoumis, Kelly. 2007. "July 2007, Theriot, C., and Tzoumis, Kelly, Bankside Chicago. Rivertown: Rethinking Urban Rivers. Pp. 65–85. MIT Press." *ResearchGate,* January, 65–85.

United Nations, Department of Economic and Social Affairs, Population Division. 2012. "World Urbanization Prospects: The 2011 Revisions."

United Nations Environmental Programme (UNEP), Ozone Secretariat. 2016. "Montreal Protocol on Substances That Deplete the Ozone Layer." UNEP Ozone Secretariat. http://ozone.unep.org/en/treaties-decisions/montreal-protocol-substances-deplete-ozone-layer.

U.S. Department of the Interior, Bureau of Reclamation. 2012. "Colorado River Basin Water Supply and Demand Study." December. https://www.usbr.gov/lc/region/programs/crbstudy.html.

U.S. Environmental Protection Agency. 2015. "National Summary of State Information WATERS US EPA." *Watershed Assessment, Tracking & Environmental Results.* Accessed February 11. http://iaspub.epa.gov/waters10/attains_nation_cy.control.

Ure, Andrew. 1831. *A Dictionary of Chemistry and Mineralogy: With Their Applications.* 4th ed. London: Thomas Tegg.

Visconti, Guido. 2014. "Anthropocene: Another Academic Invention?" *Rendiconti Lincei* 25 (3): 381–92.

Vörösmarty, C. J., P. B. McIntyre, M. O. Gessner, D. Dudgeon, A. Prusevich, P. Green, S. Glidden, et al. 2010. "Global Threats to Human Water Security and River Biodiversity." *Nature* 467 (7315): 555–61.

Vörösmarty, Charles J., and Michel Meybeck. 2004. "Responses of Continental Aquatic Systems at the Global Scale: New Paradigms, New Methods." In *Vegetation, Water, Humans and the Climate,* 375–413. Global Change—the IGBP Series. Berlin: Springer.

Vörösmarty, Charles J., Michel Meybeck, and Christopher L. Pastore. 2015. "Impair-Then-Repair: A Brief History and Global-Scale Hypothesis Regarding Human-Water Interactions in the Anthropocene." *Daedalus* 144 (3): 94–109.

Vries, Peer. 2001. "Are Coal and Colonies Really Crucial? Kenneth Pomeranz and the Great Divergence." *Journal of World History* 12 (2): 407–46.

———. 2010. "The California School and Beyond: How to Study the Great Divergence?" *History Compass* 8 (7): 730–51.

Waldman, John. 2010. "The Natural World Vanishes: How Species Cease to Matter." *Yale Environment 360.* http://e360.yale.edu/feature/the_natural_world_vanishes_how_species_cease_to_matter/2258/.

Walker, Brad. 2012. "OUR FUTURE? A Vision for a Land, Water, and Economic Ethic in the Upper Mississippi River Basin." Environmental Report. St. Louis: Missouri Coalition for the Environment.

Waller, Martyn. 1994. *The Fenland Project, Number 9: Flandrian Environmental Change in Fenland.* 70. Cambridge: Cambridgeshire Archaeological Committee. http://eprints.kingston.ac.uk/id/eprint/17568.

Wallerstein, Immanuel. 2011a. *The Modern World-System I: Capitalist Agriculture and the Origins of the European World-Economy in the Sixteenth Century.* Berkeley: University of California Press.

———. 2011b. *The Modern World-System II: Mercantilism and the Consolidation of the European World-Economy, 1600–1750.* Berkeley: University of California Press.

———. 2011c. *The Modern World-System III: The Second Era of Great Expansion of the Capitalist World-Economy, 1730s–1840s.* Berkeley: University of California Press.

———. 2011d. *The Modern World-System IV: Centrist Liberalism Triumphant, 1789–1914.* Berkeley: University of California Press.

Walter, Robert C., and Dorothy J. Merritts. 2008. "Natural Streams and the Legacy of Water-Powered Mills." *Science* 319 (5861): 299–304.

Water Footprint Calculator. 2015. "National Water Footprints." Accessed February 11. http://www.waterfootprint.org/?page=cal/waterfootprintcalculator_national.

Water Framework Directive (Directive 2000/60/EC of the European Parliament and of the Council Establishing a Framework for the Community Action in the Field of Water Policy). 2000. http://ec.europa.eu/environment/water/water-framework/index_en.html·

Waters, Colin N., Jan Zalasiewicz, Colin Summerhayes, Anthony D. Barnosky, Clément Poirier, Agnieszka Gałuszka, Alejandro Cearreta, et al. 2016. "The Anthropocene Is Functionally and Stratigraphically Distinct from the Holocene." *Science* 351 (6269): 2622.

Welky, David. 2011. *The Thousand-Year Flood: The Ohio-Mississippi Disaster of 1937.* Chicago: University of Chicago Press.

White, Gilbert Fowler. 1942. "Human Adjustment to Floods: A Geographical Approach to the Flood Problem in the United States." PhD dissertation, University of Chicago.

White, Iain. 2010. *Water and the City: Risk, Resilience and Planning for a Sustainable Future.* London: Routledge.

White, Lynn. 1962. *Medieval Technology and Social Change.* Oxford: Oxford University Press.

White, Richard. 1996. *The Organic Machine: The Remaking of the Columbia River.* Oxford: Oxford University Press.

Whitington, Jerome. 2016. "Modernist Infrastructure and the Vital Systems Security of Water: Singapore's Pluripotent Climate Futures." *Public Culture* 28 (279): 415–41.

Wiens, John A. 2002. "Riverine Landscapes: Taking Landscape Ecology into the Water." *Freshwater Biology* 47 (4): 501–15.

Wilkerson, I. 1992. "Chicago's Loop Is Closed Down as River's Water Floods Tunnels." *New York Times,* April 14.

Williams, Mark, Jan Zalasiewicz, Neil Davies, Ilaria Mazzini, Jean-Philippe Goiran, and Stephanie Kane. 2014. "Humans as the Third Evolutionary Stage of Biosphere Engineering of Rivers." *Anthropocene* 7 (September): 57–63.

Wittfogel, K. A. 1957. *Oriental Despotism: A Comparative Study of Total Power.* New Haven, CT: Yale University Press.

Wohl, Ellen. 2013. "Wilderness Is Dead: Whither Critical Zone Studies and Geomorphology in the Anthropocene?" *Anthropocene* 2 (October): 4–15.

Wong, Bin. 2000. *China Transformed: Historical Change and the Limits of European Experience.* Ithaca, NY: Cornell University Press.

World Bank. 2014. "GNI per Capita, Atlas Method (Current US$) Data. World Bank Data Catalogue." http://data.worldbank.org/indicator/NY.GNP.PCAP.CD.

World Wildlife Fund. 2014. "Living Planet Report 2014: Species and Spaces, People and Places." World Wildlife Fund. http://awsassets.panda.org/downloads/wwf_lpr2014_low_res_full_report.pdf.

Worster, Donald. 1977. *Nature's Economy: A History of Ecological Ideas.* Cambridge: Cambridge University Press.

Wright, Peter D. 2014. *Life on the Tyne: Water Trades on the Lower River Tyne in the Seventeenth and Eighteenth Centuries, a Reappraisal.* Burlington, VT: Ashgate.

Wrigley, E. A. 2010. *Energy and the English Industrial Revolution.* Cambridge: Cambridge University Press.

Yeoh, Brenda S. A. 2003. *Contesting Space in Colonial Singapore: Power Relations and the Urban Built Environment.* Singapore: National University of Singapore Press.

Zalasiewicz, J. 2015. "Epochs: Disputed Start Dates for Anthropocene." *Nature* 520 (7548): 436.

Zalasiewicz, J., Mark Williams, Alan Haywood, and Michael Ellis. 2011. "The Anthropocene: A New Epoch of Geological Time?" *Philosophical Transactions of the Royal Society A: Mathematical, Physical and Engineering Sciences* 369 (1938): 835–41.

Zalasiewicz, J., M. Williams, A. Smith, T. L. Barry, A. L. Coe, P. R. Bown, P. Brenchley, et al. 2008. "Are We Now Living in the Anthropocene?" *GSA Today* 18 (2): 4–8.

Zalasiewicz, J., Mark Williams, Will Steffen, and Paul Crutzen. 2010. "The New World of the Anthropocene." *Environmental Science & Technology* 44 (7): 2228–31.

CONTRIBUTORS

JEFFREY BENJAMIN is a PhD student in archaeology at Columbia University in New York. He is an artist, writer, and art-chaeologist and lives in the Catskill Mountains of New York State.
jlb2289@columbia.edu

HELEN BERRY is Professor of History and Dean of Postgraduate Studies in the Faculty of Humanities and Social Studies at Newcastle University, U.K. She has published widely on the global interconnections that were fostered in an era of burgeoning consumerism between 1700 and 1850 and has ongoing research interests in using microhistory to make sense of "big data" relating to climate change. She is a member of the Anthropocene Research Group and the McCord Centre for Landscape Studies at Newcastle University.
helen.berry@ncl.ac.uk
School of History, Classics and Archaeology, Newcastle University, Armstrong Building, Newcastle upon Tyne NE1 7RU, United Kingdom

CELIA DEANE-DRUMMOND is Professor in Theology at the University of Notre Dame and Director of the Center for Theology, Science, and Human Flourishing. She was editor of the journal *Ecotheology* for six years and has served as chair of the European Forum for the Study of Religion and Environment from 2011 to present. Her recent books include *Ecotheology* (2008), *Christ and Evolution* (2009), *Creaturely Theology,* ed. with David Clough (2009), *Religion and Ecology in the Public Sphere,* ed. with Heinrich Bedford-Strohm (2011), *Animals as Religious Subjects,* ed. with Rebecca Artinian Kaiser and David Clough (2013), *The Wisdom of the Liminals* (2014), *Re-Imaging the Divine Image* (2014), *Technofutures, Nature and the Sacred,* ed. with Sigurd Bergmann and Bronislaw Szerszynski (2015); *Religion in the Anthropocene,* ed. with Sigurd Bergmann and Markus Vogt (2017).
Celia.Deane-Drummond.1@nd.edu
130 Malloy Hall, Department of Theology, University of Notre Dame, Notre Dame, IN 46530

MATT EDGEWORTH, archaeologist, is Senior Project Officer for the Cambridge University Archaeology Unit, and honorary research fellow at the University of Leicester, U.K. He has excavated sites in Carthage in North Africa, the Orkney Islands in Scotland, and many sites in between. He is the author of *Fluid Pasts: Archaeology of Flow*. As a member of the Anthropocene Working Group, he argues for the broader relevance of archaeological strata and artifacts in interdisciplinary debate.
me421@cam.ac.uk
Cambridge University Archaeological Unit, Department of Archaeology, University of Cambridge, Downing Street, Cambridge CB2 3DZ, United Kingdom

DAVID GILVEAR is Professor of River Science, Catchment and River Science Research Group, Sustainable Earth Institute, at Plymouth University, U.K. He is president of the International Society of River Science. Gilvear is particularly interested in interdisciplinarity and the application of his work to real-world river issues. In recent years he has been interested in the utility of the ecosystem services approach to catchments and rivers management.
david.gilvear@plymouth.ac.uk
Sustainable Earth Institute, Plymouth University, Drakes Circus PL4 8AA, United Kingdom

STEPHANIE C. KANE is Professor in the Department of International Studies, Indiana University, Bloomington. She is an ethnographer of water in cities. Her book, *Where Rivers Meet the Sea: The Political Ecology of Water* (Temple, 2012) is about Salvador da Bahia and Buenos Aires. Her essays have appeared in numerous journals, including *Human Organization, Social Text, Ethnobiological Letters, Political and Legal Anthropology, Journal of Folklore Research, Crime Media Culture,* and *Signs,* and edited volumes, including most recently, *Anthropology of Environmental Health* (2016), *Reflexivity in Criminological Research* (2014), *International Handbook of Green Criminology* (2013), *Comparative Decision Making* (2013), and *Transforming Urban Waterfronts* (2011).
stkane@indiana.edu
Department of International Studies, School of Global and International Studies E1001, Indiana University, Bloomington, IN 47405

JASON M. KELLY is Director of the IUPUI Arts and Humanities Institute (IAHI) and Associate Professor of History in the Indiana University School of Liberal Arts at IUPUI. He is the author of *The Society of Dilettanti* (Yale University Press and the Paul Mellon Centre for Studies in British Art, 2010) and has published articles in the *Journal of British Studies,* the *British Art Journal,* the *Walpole Society,* and more. His research explores the historical intersections of art, science, and philosophy through studies of the Enlightenment. He currently leads the Rivers of the Anthropocene project, which brings together an international team of scientists, humanists, artists, policy makers, and community members to study global river systems since 1750. As the inaugural director of the IAHI, he creates and fosters transdisciplinary research collaborations that have deep engagements with local communities. Kelly is a Fellow of the Society of Antiquaries of London, the 2013 recipient of IUPUI's Research Trailblazer Award, and a two-time recipient of the IU Trustees Teaching Award.
jaskelly@iupui.edu
IUPUI Arts & Humanities Institute, 755 W. Michigan St., UL 4115T, Indianapolis, IN 46202

ANDY LARGE is a reader in River Science and convenor of the Anthropocene Research Group in the School of Geography, Politics and Sociology, Newcastle University. He is a field-based physical geographer with almost thirty years of experience in researching river catchment systems. His research focuses on improving our understanding of how extreme events drive responses to river flooding and to what extent this is translated in societal terms as "geohazards." Large publishes on innovative ways of digitally sensing rivers, and as convenor of the Anthropocene Research Group at Newcastle University he is especially interested in developing innovative ways of bringing disciplines together to look at rivers in new and exciting ways.

School of Geography, Politics and Sociology, Newcastle University, 5th Floor Claremont Tower, Newcastle upon Tyne NE1 7RU, United Kingdom

LAURENCE LESTEL is a researcher at the Centre National de la Recherche Scientifique. She defended her PhD in chemistry in 1987 and, in 2003, an Habilitation in the historical approach to contemporary environmental issues. She develops "environmental history" in two directions: the social history of the relationship between humans and their environment and the history of water contamination. She works in collaboration with biogeochemists of the METIS laboratory (University Paris 6, Pierre and Marie Curie) and the interdisciplinary program on the Seine River and its basin (PIREN Seine), where she has focused on metal contamination over the past 150 years. She directed an international comparison of the relationships between four European cities and their rivers since 1870 (Lestel and Carré, *Les rivières urbaines et leur pollution,* in press at Indisciplines, Quae editions, Paris, 2017). laurence.lestel@upmc.fr

Sorbonne Universités, UPMC Univ Paris 06, CNRS, EPHE, UMR 7619 Metis, Paris, France

KEN LUBINSKI is recently retired from his position as Chief of River Ecology, U.S. Geological Survey, Upper Midwest Environmental Sciences Center, La Crosse, Wisconsin. He began his studies of the Illinois and Mississippi Rivers in the early 1970s. His primary interest is the development of science-based adaptive management processes and the communication of river knowledge to nonscientific audiences. He remains a board member of the International Society for River Science, advises the Fishers and Farmers National (U.S.) Fish Habitat Partnership, and coordinates a floodplain forest restoration effort in the Reno Bottoms of the Mississippi River. kenlubinski@outlook.com

SINA MARX holds a degree with honors in anthropology, political science, and Indonesian studies. She has worked on water and development in different parts of the world, for example, for the Deutsche Gesellschaft für Internationale Zusammenarbeit (GIZ), the Center for Development Research (ZEF), and the Global Water System Project (GWSP). She is currently working both as the head office coordinator of the German Committee for Disaster Reduction in Bonn and as a consultant for project management and public relations on development and sustainability. sina.marx@gmail.com

MICHEL MEYBECK is an emeritus scientist at the French Centre National de la Recherche Scientifique, within the Metis laboratory (University Paris 6, Pierre and Marie Curie). As a geochemist he first worked on the natural composition of water and particulates in rivers, determining river-borne fluxes to oceans and their control factors. He has been the scientific advisor (1978–98) of the UNEP-WHO-UNESCO global network of water quality monitoring and assessment (GEMS-Water). He combined these two approaches to river basins with the International Geosphere-Biosphere Programme (IGBP, 1995–2005) and, since 1989, within the PIREN Seine program, which studies complex human and river interactions.

michel.meybeck@upmc.fr
Sorbonne Universités, UPMC Univ Paris 06, CNRS, EPHE, UMR 7619 Metis, Paris, France

MARY MISS is the founder of City as Living Laboratory, a framework for making issues of sustainability tangible through collaboration and the arts. She has reshaped the boundaries between sculpture, architecture, landscape design, and installation art by articulating a vision of the public sphere where it is possible for an artist to address the issues of our time. She has worked with architects, planners, engineers, ecologists, and public administrators on projects as diverse as creating a temporary memorial around the perimeter of Ground Zero and marking predicted flood levels of Boulder, Colorado.

studio@marymiss.com

PHILIP V. SCARPINO earned his PhD in environmental history at the University of Missouri, Columbia, in 1983. He is Professor of History and Director of the Graduate Program in Public History at IUPUI. Scarpino is a cofounder of the Rivers of the Anthropocene Project. Selected, related publications include *Great River: An Environmental History of the Upper Mississippi River* (1985); "Interpreting Environmental Themes in Exhibit Format," in *Public and Environmental History,* coedited by Martin Melosi and Philip Scarpino (2004); "Large Floodplain Rivers as Human Artifacts: A Historical Perspective on Ecological Integrity," U.S. Geological Survey, Special Refereed Report (1997); "Great Lakes Fisheries: International Response to the Decline of the Fisheries and the Lamprey/Alewife Invasion," in *A History of Water,* vol. 2, *The Political Economy of Water,* ed. Terje Tvedt and Richard Coope (2006); "Addressing Cross-Border Pollution of the Great Lakes," in *Transnationalism in Canada–United States History into the Twenty First Century,* ed. Michael Behiels and Reginald Stuart (2010); and "A Historian's Perspective on Rivers of the Anthropocene," in *The Global Water System in the Anthropocene: Challenges for Science and Governance,* ed. Janos Bogardi et al. (2014).

DINAH SMITH is Honorary Visiting Fellow in the Department of Geology at the University of Leicester, U.K. In 2002, after thirty-four years of teaching in primary schools, Smith undertook studies at the University of Leicester in the geology department and was later awarded an MGeol. This led to research connected with the Fenland of East Anglia, England, and the roddons, fossilized creek systems of the Holocene. She received her PhD in 2014. She still maintains many connections with the Fenland area.

dms23@le.ac.uk
Department of Geology, University of Leicester, Leicester LE1 7RH, United Kingdom

ELEANOR STARKEY is a PhD researcher in Catchment Science in the School of Civil Engineering and Geosciences at Newcastle University and has a Master's degree in hydrology and climate change. Her work is focused on understanding river behavior, low-cost data collection methods, managing floods, engaging with communities, and the emerging discipline of citizen science. Starkey's work with small communities has been of significance on a national scale, allowing scientists and engineers to appreciate the importance of working closely with the public and to capture and utilize local knowledge.

eleanor.starkey@ncl.ac.uk

School of Geography, Politics and Sociology, Newcastle University, 5th Floor Claremont Tower, Newcastle upon Tyne NE1 7RU, United Kingdom

JAMES SYVITSKI is Executive Director of the Community Surface Dynamics Modeling System and Professor at University of Colorado, Boulder. He received doctorate degrees in oceanography and geological science from the University of British Columbia in 1978 and held various appointments within Canadian universities (1978–95) while a senior research scientist at the Bedford Institute of Oceanography. Syvitski was director of INSTAAR from 1995 to 2007 and holds University of Colorado faculty appointments in geological sciences, applied mathematics, atmosphere & ocean sciences, hydrological sciences, and geophysics. Since 2007, he has been the executive director of CSDMS, an international effort to develop, support, and disseminate integrated computer models to the broader geoscience community. He chaired the International Geosphere-Biosphere Programme (2011–16), an effort to guide society onto a sustainable pathway during rapid global change.

james.syvitski@colorado.edu
CSDMS, Sustainability, Energy and Environment Complex (SEEC), University of Colorado, 4001 Discovery Dr., Boulder, CO 80303

MARTIN THOMS is Professor of River Science and Director of the Institute for Rural Futures at the University of New England, Armidale, NSW, Australia. His research focuses on the intersection between the disciplines of hydrology, geomorphology, and ecology and the science-management-policy boundary. He is past president of the International Society for River Science, an interdisciplinary society that considers rivers as true social-ecological systems.

Martin.Thoms@une.edu.au
Director, Institute for Rural Futures, University of New England, Armidale NSW 2351, Australia

MARK WILLIAMS is Professor of Palaeobiology in the Department of Geology at the University of Leicester, U.K. He is interested in major events in the four-billion-year history of life and their relevance for interpreting human-induced changes to the biosphere. With Jan Zalasiewicz, he is coauthor of the books *Ocean Worlds* and *The Goldilocks Planet,* and they are currently working on a new book examining the role of skeletons in building the complexity of the biosphere.

mri@le.ac.uk
Department of Geology, University of Leicester, Leicester LE1 7RH, United Kingdom

JAN ZALASIEWICZ is Professor of Palaeobiology in the Department of Geology at University of Leicester, U.K. He is a geologist as well as a palaeontologist and formerly worked for the British Geological Survey. He has broad interests in sedimentology, stratigraphy, paleontology, and paleoenvironmental analysis. He currently chairs the Anthropocene Working Group and is also involved with other stratigraphic bodies both nationally and internationally. His books include *The Earth after Us, The Planet in a Pebble,* and *Rocks: A Very Short Introduction,* as well as titles coauthored with Mark Williams.

jaz1@le.ac.uk
Department of Geology, University of Leicester, Leicester LE1 7RH, United Kingdom

INDEX